Accessing...

Science

Teacher Resource Book

Year 5

Christine Moorcroft

Acknowledgements

United Kingdom: Folens Publishers, Apex Business Centre, Boscombe Road, Dunstable, LU5 4RL
Email: folens@folens.com

Ireland: Folens Publishers, Greenhills Road, Tallaght, Dublin 24
Email: info@folens.ie

Poland: JUKA, ul. Renesansowa 38, Warsaw 01-905

Editor: Chris Affleck
Layout artist: Patricia Hollingsworth
Illustrations: Alan Baker and Paul Stanway
Cover design: Martin Cross
Cover image: CORBIS/Jim Sugar

First published 2005 by Folens Limited.
Every effort has been made to contact copyright holders of material used in this publication. If any copyright holder has been overlooked, we should be pleased to make any necessary arrangements.

British Library Cataloguing in Publication Data. A catalogue record for this publication is available from the British Library.

ISBN 1 84303 564 2

Contents

Introduction

Folens ***Accessing... Science Books*** provide a collection of visual resources around which teachers can plan lessons based on the QCA schemes of work for the Foundation Stage and Key Stages 1 and 2.

Accessing... Science Books are a unique resource, which can provide every child in the class with high-quality visual material. It can be used as a starting point for whole-class or group discussions, or for individual work.

The ***Big Books*** and pupil ***Photo Books*** contain no text apart from captions and, occasionally, images which contain text. This makes the series a versatile resource which can be used by children with very different reading abilities. It also allows the maximum use of space for useful source material.

Accessing... Science Books are intended to be used by teachers to supplement their school's scheme of work rather than as a complete scheme in themselves.

Big Books and pupil ***Photo Books*** for **Science** include photographs and illustrations of items which the children might not otherwise have the opportunity to see; although their learning will be enriched if they also have the chance to experience some of these: people, animals, plants, places, materials, scientific phenomena and even scientists and their experiments and investigations.

The use of ***Big Books*** and multiple copies of ***Photo Books*** enables whole classes or groups of children to use the pictures to help them to complete tasks suggested in the ***Teacher Resource Books***.

The ***Teacher Resource Books*** are designed to help busy teachers, who cannot be expected to be experts in every subject, to prepare interesting and motivating science lessons. They provide:

- ❑ background information for teachers
- ❑ vocabulary lists
- ❑ starting points
- ❑ suggestions for individual and group discussions and activities based on the visual resources
- ❑ discussion points for plenary sessions
- ❑ ideas for extensions for able children and support for less able children.

Additional activities are provided in the photocopiable pages.

Foundation Stage and Key Stage 1
The material for each of these age groups comprises:

Foundation Stage and Year 1
- ❑ a 20-page ***Big Book***
- ❑ four A4 card copies of each poster from the ***Big Book*** which can be used flexibly: for example, for display or for groups or individuals to use as a stimulus for written or practical work or for reference
- ❑ a ***Teacher Resource Book*** supplying ideas for lessons and photocopiable activity sheets providing information, stories and activities

Year 2
- ❑ a 24-page ***Big Book***
- ❑ a 48-page ***Photo Book***
- ❑ a 64-page ***Teacher Resource Book***

Key Stage 2
The materials for these age groups comprise:

- ❑ a 48-page ***Photo Book*** for each year group
- ❑ a 64-page ***Teacher Resource Book*** for each year group

Accessing... Science presents the materials and information which teachers need in order to follow the QCA schemes of work for Science and the Curriculum for the Foundation Stage (Knowledge and understanding of the world). Authors who know the subjects have done the necessary background research for the teacher.

Keeping healthy

1 'The Potato Eaters' by Víncent van Gogh

Photo Book reference Page 4

QCA links
The children learn that we need an adequate and varied diet in order to stay healthy. (QCA Unit 5A)

Background information

Miners and their families in the Borinage region of Belgium – painted by Vincent van Gogh in 1885 – survived on a diet in which there was little variety. Their staple food was potatoes; they were grown locally and were plentiful and cheap. Potatoes are high in fibre and contain no fat. They provide carbohydrates (mainly starch), vitamin C, a little protein and potassium and small amounts of iron. A potato weighing 100g provides 66 calories.

Starting points

- Review the children's previous learning about grouping foods according to their main nutrients and how the body uses them: protein for growth, fat for stored energy, carbohydrates for immediate energy, fibre for a healthy digestive system and vitamins and minerals to help the body to make the best use of foods.
- Discuss whether potatoes are a healthy food (a food we should eat often) and ask the children to support their answers with reasons.
- *Do the children eat potatoes at home or at school?* Ask them how they prefer potatoes to be served: for example, boiled, roasted, baked, fried, mashed, as chips or crisps. *Which do they think are the healthiest ways of preparing potatoes?*

Using the picture

- Discuss the meal shown in the picture and point out that poor mining families in 1855 in the Borinage, Belgium, lived mainly on potatoes because they were grown locally and were cheap. This family is drinking coffee (without milk or sugar). Ask the children if they are likely to be healthy if they live mainly on potatoes and coffee.
- Challenge the children to find out about the main nutrients the people in this family are taking in during this meal. *In which nutrients might their diet be deficient?*

Activities

- Ask the children to use different sources to find out about the nutrients in potatoes: for example, leaflets from health centres, supermarkets and pharmacies, the Internet (see **Resources**) and information books. They could also find out about the recommended daily amounts of these nutrients for people of different ages.
- Ask them to list any nutrients they think might be missing from the diets of the 'Potato Eaters'.
- The children could also contribute to a 'potato recipe' book: they could collect recipes from cookery books or write to celebrity chefs for their favourite potato recipes. If possible, try them out in school.
- Foods can be tested for starch by dropping a small amount of tincture of iodine on to them. If starch is present the food turns dark blue.

> ⚠ An adult should do this.

Higher-achieving children
Challenge the children to find out which other foods could be eaten, along with potatoes, for a healthy diet.

Lower-achieving children
Provide clues to help them to suggest foods to complement the Potato Eaters' diet: *What could they eat with the potatoes for fat? What could they eat with them for protein? What else could they eat, for vitamins and fibre?*

Plenary session

- Explain that, although people can survive on a diet of mainly one type of food, they are healthier if they eat a variety of foods.

Vocabulary

carbohydrate, deficiency, deficient, fat, iron, mineral, nutrient, potassium, protein, starch, sugar, vitamin

Photo Book reference Page 5

2 James Lind

QCA links
The children learn how scientific ideas about diet can be tested. (QCA Unit 5A)

Background information

Scurvy is a disease caused by a deficiency of vitamin C and was once very common among sailors because of the lack of fresh fruit in their diet. The deficiency causes a breakdown of collagen – a protein which binds cells together – resulting in the breakdown of the walls of blood capillaries. Symptoms include weakness, black-and-blue marks on the skin (caused by internal haemorrhaging), pain in the joints, bleeding gums, loose teeth and painful raised red spots around hair follicles. James Lind was a Scottish physician at a Royal Navy hospital who specialised in the prevention and treatment of the diseases which were common among sailors in the eighteenth century.

Starting point

- Tell the children about the symptoms of scurvy and that, in the past, the disease was much more common among sailors than other groups of people. *If they were scientists at the time what might they have listed as some of the possible causes of the disease?* (Focus on how the lifestyles of sailors might have differed from those of other people.)

Using the pictures

- Ask the children what they can find out about James Lind from the picture: for example, when he lived, the type of work he did, and where, his interests and why he became famous. Encourage them to use as many clues from the picture as possible and then to use other sources of information.

Activities

- Ask the children to find out about the differences between the lives of sailors and other people. They could record these on a chart:

Differences	
Life at home	Life at sea

- *If the children had been scientists at the time, how would they use this information to find out what caused scurvy?*
- After the children have completed the chart, ask them what differences they identified between life at home and life at sea. Draw out that the diet of sailors would have included very little fresh food because ships had no refrigerators in those days and voyages took much longer than they do nowadays. Point out that people in the eighteenth century knew a lot less about the effects of diet than we do now. Introduce the idea that diseases can be caused by poor diet.

Higher-achieving children
Ask them to research the nutrients provided by fresh fruit and vegetables and to suggest which of these is important in preventing scurvy.

Lower-achieving children
Provide information about vitamin C and explain that this is important in preventing scurvy but that this was not known in the eighteenth century.

Plenary session

- Draw out that the lack of a nutrient in the diet can affect people's health and even lead to deficiency diseases.

Vocabulary

bleeding, blood vessels, capillaries, citrus, deficiency, deficient, vitamin

2 James Lind's experiment

Photo Book reference Page 6

QCA links
The children learn how scientific ideas about diet can be tested. (QCA Unit 5A)

Background information

See also topic 2. James Lind observed that citrus fruits contained something which counteracted the effects of scurvy; he carried out experiments which resulted in his *Treatise on the Scurvy* (1753) and later developed a method of concentrating and preserving citrus juices for use at sea. In 1795, the British Royal Navy began to provide a daily ration of lime or lemon juice for all seafarers – leading to the nickname 'limeys'. At that time vitamins had not been discovered.

Starting point

■ The children should first have completed topic 2. Draw out the main points which they learned about how scientists can begin to find the causes of an illness. Tell them that people began to notice that some sailors did not suffer from scurvy and that others could be cured. They thought that certain foods or medicines might help, but they were not sure which ones.

Using the pictures

❑ Ask the children to read about James Lind's experiment. *What did he observe? What question did this make him ask? How did he make the test fair? What did he expect to happen? Did the results of the experiment support his idea?*

❑ Discuss how reliable the experiment was: for example, the number of people tested and the fairness of the test.

❑ Ask the children why they think going to sea should have caused a disease caused by a lack of fruit and vegetables in the diet. Explain that ships had to stock up before a voyage and often sailed for a long time before they could collect fresh food; a deficiency of vitamin C causes scurvy and vitamin C, unlike some other vitamins, cannot be stored or manufactured by the body.

Activities

- *Can the children name any citrus fruits?* Draw out that citrus fruits are good sources of vitamin C and ask the children to collect and label pictures of, and information about, as many different citrus fruits as possible for a display. The children could find out why these fruits are named *citrus* fruits.
- Ask them to find out about James Lind's work on preserving citrus fruit juices and how his work helped the Royal Navy. They could write a letter, as if from James Lind to the Royal Navy officials, explaining how to prevent sailors from suffering from scurvy at sea and drawing out the benefits of this.

Higher-achieving children
Ask them to research another deficiency disease and to produce a leaflet about how to avoid it. They could present their leaflet using a word-processor and by importing pictures.

Lower-achieving children
Provide a structure to help them to write a letter 'from James Lind' to an admiral in the Royal Navy. Provide an introductory paragraph and then ask them to add the information about the experiment and the conclusions which were drawn from it.

Plenary session

◆ Ask the children what they have learned about how scientists turn their observations into questions which can be investigated and how they plan and carry out experiments.

Vocabulary

bleeding, blood vessels, capillaries, citrus, deficiency, deficient, vitamin

Photo Book reference Page 7

4 Exercise

QCA links
The children learn that we need exercise in order to stay healthy. (QCA Unit 5A)

Background information

Stamina means the ability to continue with a strenuous activity for a long time; *strength* means the ability of the muscles to support the weight of the body and additional weights; *suppleness* means the mobility of the joints between the bones; *co-ordination* means the way in which the body uses the muscles to respond to what is observed (for example, catching, kicking or batting a ball).

Starting point

■ Ask the children to give examples of exercise. Write up their responses and ask them if they can give a definition of exercise. Draw out that exercise involves movement; when people exercise they move their muscles and their bones.

Using the pictures

❑ Discuss what is meant by the captions of the photographs (see Background information) and ask the children to select other exercises from the list they suggested (see Starting point) which develop stamina, strength, suppleness and co-ordination. Ask: *Which sport shown here uses the most energy? How can you tell from the photographs?*

❑ They could record their ideas on a chart:

Exercises for different purposes			
Stamina	Strength	Suppleness	Co-ordination

❑ Point out that some exercises can develop more than one of these attributes.

Activities

- During a PE lesson, try out different forms of exercise with the children and ask them which parts of their bodies they are exercising.
- *Which exercises make their heart and lungs work hard? How can they tell?* Draw out that they might be able to feel their heart beating more quickly and they might breathe more quickly.
- Show the children video clips of sporting events and ask them to notice which parts of their bodies the athletes are exercising. Include ice skating, gymnastics football, table tennis, tennis, running, swimming, javelin and discus.

Higher-achieving children
They could do some research, using information books, sports leaflets and the Internet, to find out how much energy each sport uses in fifteen minutes.

Lower-achieving children
Provide a chart on which they can record which parts of the body are exercised by different sports:

	Parts of the body exercised (✓)				
Sport	Arms	Legs	Back	Heart	Lungs

Plenary session

◆ Ask the children how exercise helps to keep the body healthy.

Vocabulary

co-ordination, exercise, joint, muscle, stamina, strength, suppleness

5 The heart

Photo Book reference Page 8

QCA links
The children learn that the muscle in the walls of the heart contracts regularly, pumping blood around the body, and that blood vessels carry blood around the body. (QCA Unit 5A)

Background information

The process by which energy is used to enable a body to keep warm, carry out essential processes or move is similar to burning; it uses up oxygen and a fuel, gives off carbon dioxide and produces heat and other energy. Blood carries oxygen around the body to enable these processes to take place. A pump (the heart) is required to move the blood around the body through the blood vessels (arteries, veins and capillaries). Oxygen enters the blood through the walls of fine blood vessels around the lungs; it enters the heart and is pumped around the body. Blood whose oxygen has been used up and replaced with carbon dioxide is pumped back to the heart and on to the lungs to collect more oxygen. We breathe out carbon dioxide, which has passed from the blood into the lungs. A heartbeat is the sound made by the heart valves closing; the normal heart rate for an adult is about 70 beats per minute. It is slightly faster for children.

Starting points

- Invite volunteers to come out and draw a picture of a human heart. Discuss the shape of the heart. *Did they draw a regular shape as depicted on playing cards? Is the heart really this shape?*
- *Can the children point out the location of their heart?*
- *Do they know what it is for?*

Using the pictures

- Ask the children to compare their drawings of hearts (see Starting points) with these pictures. Point out that the heart is located in the centre of the chest, close to the lungs, and ask the children what protects it (the ribs).
- Ask them to make notes about what they can find out from the pictures about the purpose of the heart and how it works, including the meaning of 'heartbeat'. They could use their notes to help them to write an explanation of how the heart works.

Activities

- Ask the children if they can feel their heart beating. They could use a stethoscope to listen to one another's heartbeats and make sound recordings of their heartbeats using a cassette recorder with the microphone pressed on to their chest.
- During a PE lesson, ask the children to move around quickly: they could run around the playground, run on the spot, skip, hop, jump, and so on.
- *How did exercise affect their heartbeat? How can they tell?* They could make sound recordings of their heartbeat before and after exercise, play them back, with the volume as loud as is necessary, and count the heartbeats over 30 seconds. Ask them to record their findings on bar charts (see ***Photocopiable sheet 1***).
- ***Photocopiable sheet 1*** provides formats for bar charts on which the children can record their heart rate and pulse rate before and after exercise (see also topic 7).

Higher-achieving children
Ask them to use their own results, and those of others, to help them to work out the average number of heartbeats per minute at rest and after strenuous exercise.

Lower-achieving children
Encourage them to look at the lengths of the columns they mark on the bar chart to find answers to questions such as, *How many times did your heart beat in half a minute before exercise? How many times did it beat in half a minute after exercise? What does exercise do to your heartbeat?*

Plenary session

- Encourage the children to draw conclusions from their investigations about how exercise affects the heartbeat. Discuss why the heart might need to beat faster after exercise. Ask: *What is it doing when it beats?* (Pumping blood around the body.)

Vocabulary

air, blood, carbon dioxide, gas, heart, heartbeat, lungs, oxygen, pump, stethoscope

6 Measuring blood pressure

Photo Book reference Page 9

QCA links
The children learn that the muscle in the walls of the heart contracts regularly, pumping blood around the body, and that blood vessels carry blood around the body. (QCA Unit 5A)

Background information

The heart pumps blood around the body through the arteries. Each time the heart pumps it forces blood through the arteries and into capillaries (smaller blood vessels). ***Blood pressure*** is the term for the force produced by the heart in the arteries. Blood pressure increases when the heart contracts to force blood through the arteries (systolic pressure) and reduces when the heart relaxes (diastolic pressure). The photograph shows a sphygmomanometer being used to measure blood pressure; this is a device which uses the level of mercury in a tube to give readings of the systolic and diastolic pressure. A normal blood pressure is about 130/80 (often referred to as '130 over 80' – a systolic pressure of 130 and a diastolic pressure of 80).

Starting point

■ Ask the children if they have heard the term *blood pressure*. They might have heard family members talking about having their blood pressure measured or have come across health campaigns aimed at helping people to reduce their blood pressure. Point out that everyone has blood pressure – it is the force of the blood being pumped around the blood vessels by the heart. Remind the children about their previous learning about the heart. You could mention that some people develop high blood pressure, and that this is unhealthy; it means that the heart has to work extra hard in order to pump blood around the body.

Using the picture

❑ Ask the children to describe what is happening in the photograph. Ask: *What might the technician be able to hear? Do they know the name of the instrument she is using for listening to the pulse beat?* Explain that every time the heart beats there is a rush of blood along the arteries, away from the heart. This can be felt (and heard) as a beat or throb (*pulse*) in places such as the crook of the elbow. The technician listens to the pulse and notes the reading on the sphygmomanometer (the device attached to the band around the patient's arm). Explain that it measures the force of the heart pumping blood around the blood vessels.

Activities

- Provide some plastic tubing and a bowl of water and let the children investigate how to fill the tubing with water. Discuss what they need to do to force the water through the tubing. They could fill a syringe with water and attach it to a piece of plastic tubing (use tubing whose diameter gives a snug fit over the nozzle of the syringe). Ask them how they can force the water through the tubing. Explain that they used the syringe as a pump and that the heart has to pump blood around the body by forcing it through the tubes known as blood vessels.
- Ask them to write a report about their investigation.

Higher-achieving children
They could write an explanation of blood pressure using their investigation as a model.

Lower-achieving children
Provide a format to help them to write a report about their investigation, for example:

What we were trying to do.	How well it worked.
What we did to improve the results.	How well it worked.
Why it worked.	

Plenary session

◆ Draw out that the heart is a pump which forces blood through the blood vessels so that blood is taken around the body.

Vocabulary

blood, blood vessel, heart, pressure, pulse, pump, sphygmomanometer

Photo Book reference Page 10

QCA links

The children learn how to measure their pulse rate and to relate it to heart rate and to repeat their measurements to check their accuracy. They identify factors which could affect pulse rate, make predictions about the changes and plan what evidence to collect in order to check their predictions. (QCA Unit 5A)

7 Measuring pulse beat

Background information

Each time the heart beats there is a rush of blood along the arteries, away from the heart. This can be felt as a beat or throb (pulse) in places such as the back of the knee, the side of the neck, the wrist and the crook of the elbow.

Starting point

- Ask the children if they know what *pulse* means. *How is it linked to heartbeat?* (See Background information). *Where can they feel their pulse?*

> Warn the children not to press on the pulse spots, but to rest their fingers on them lightly. If they cannot feel the pulse in one of the places shown, they should try another. Different people will find it easier to feel their pulse in different places.

Using the pictures

- Ask the children to use the pictures to help them to find places where they can feel their pulse beat. They could count the number of pulse beats while a partner times 30 seconds. Ask them to repeat this to check it for accuracy. Ask: *Is it the same in different parts of the body?* Discuss why it should be the same.
- Ask them to compare this with their heart rate (see topic 5).
- *Do they think their pulse beat stays the same at all times, or does it change?* Discuss the factors which might make it change.

Activities

- The children could measure their pulse rate after different activities: for example, walking around the playground at a brisk pace for 5 minutes, running on the spot for 5 minutes, skipping for 5 minutes, or other activities they think of. Before each activity they should rest until their pulse rate has returned to its normal rate (see Using the pictures).
- ***Photocopiable sheet 1*** supports this and topic 5. The children should first have listened to, counted and recorded the number of times their hearts beat in 30 seconds. After they have counted their pulse beat and recorded it on the bar chart, ask them what they notice about their heart and pulse rates and discuss why they should be the same. Explain *pulse beat* (see Background information).

Higher-achieving children

Point out that the heart beats faster to send the blood around the body more quickly. Ask the children why they think exercise affects the pulse (and heart) rate. Ask: *Why should the blood need to be pumped around the body more quickly after exercise?* (Remind them about their previous learning about the heart and lungs from topic 5). They could find out, and write an explanation.

Lower-achieving children

Before they begin this activity provide practice in using a stopwatch (setting it and interpreting the readings).

Plenary session

- Draw out that each time the heart beats, it can be felt as a pulse in certain parts of the body and that we can count these beats in order to measure the heart rate.

Vocabulary

beat, blood, heart, heartbeat, pressure, pulse, pump, rate

8 Fruits from flowering plants

Photo Book reference Page 11

QCA links
The children learn that flowering plants reproduce and that many fruits and seeds provide food for animals, including humans. They make careful observations of seeds. (QCA Unit 5B)

Background information

The seeds of some flowering plants are fleshy or are surrounded by fleshy material, which is eaten by different animals. This does not necessarily destroy the seeds. Many seeds can pass through the digestive system of animals, and are dispersed as the animals move away from the source plant. Some seeds germinate more readily after passing through the digestive system of an animal.

Starting point

> Warn the children that the seeds of some plants are poisonous (see **Resources** for sources of information about poisonous plants).

■ Ask the children to name some fruits. Ask: *What is a fruit?* Ask them to consider the fruits they have named and then to say how they are similar. Draw out that fruits contain seeds or are, themselves, fleshy seeds. *Do the children know why plants produce seeds?*

Using the picture

❑ Discuss the similarities and differences between these fruits: pears, sloes and acorns grow on trees, but cotoneaster is a bush. The difference between a bush and a tree is mainly in their size. The blackthorn, on which sloes grow, is sometimes referred to as a bush, although it can grow to more than 6 metres in height. Explain that all fruits grow from the flowers of the plant. The sizes, colours and textures of the fruits are different and only two of them can be eaten by humans (pear and sloe, although sloes have a very bitter taste which makes them suitable only for flavouring drinks and preserves).

❑ Ask the children to list other fruits which are similar to each of these and compile a table of four columns – one for each type of fruit. They could classify them according to their seeds: a single stone surrounded by flesh (sloe, damson, plum, apricot, peach, nectarine); a hard nut-like seed (acorn, hazel, horse chestnut); a berry (cotoneaster, rowan, hawthorn); a fleshy fruit containing several pips (pear, apple).

Activities

- Encourage the children to bring in the fruits of plants. They could cut them open, examine them (using magnifying glasses) and make observational drawings, which could be labelled with the help of information books and electronic sources.

> Warn the children to keep their hands away from their faces when they are working, and to wash their hands after handling fruits and seeds.

- Ask them to find out at what time of the year each plant produces fruit (mainly autumn) and at what time of the year the seeds germinate (mainly spring). *What happens to the fruits and seeds during the winter?* Fruits which have not been eaten by animals rot (micro-organisms cause rotting) and the seeds lie dormant until the weather becomes warmer. Many seeds require a period of cold before they can germinate: for example, rowan.

Higher-achieving children
They could write a leaflet to help others to identify poisonous seeds and fruits.

Lower-achieving children
Help them to match the seeds and fruits to pictures to identify them.

Plenary session

◆ Draw out that plants produce seeds in order to reproduce and that some seeds are part of a fruit which can be eaten by animals.

Vocabulary

acorn, berry, fruit, pear, poisonous, reproduce, seed, sloe

Photo Book reference Page 12

9 Seed dispersal

QCA links
The children learn that plants have different ways of dispersing their seeds. (QCA Unit 5B)

Background information

Seeds scattered by the wind include dandelion, thistle, sycamore, ash, cedar, columbine, fir; seeds dispersed through the digestive systems of animals include strawberry, raspberry, cotoneaster, fig, apple, beech, redcurrant, rowan; seeds dispersed on the coats of animals include burdock, teasel, various grasses; seeds scattered when the plant is shaken include poppy, grasses; seeds dispersed by explosion include meadow cranesbill, laburnum, lupin and seeds dispersed by floating on water include sea rocket, coconut, yellow iris.

Starting points

- Ask the children to imagine all the seeds from one plant (for example, a dandelion or poppy) landing on the soil around the plant. Ask: *What would happen when they germinated? What might make it difficult for them all to survive?* Discuss how plants spread their seeds farther afield so that they are not competing for the same small patch of soil and, if the parent plant is a *perennial* (explain this term), they do not grow in the shadow of the larger parent plant and compete for space in soil which is already taken up by its roots.
- Introduce the term *dispersal* for the spreading of seeds and point out that plants can disperse their seeds in different ways.

Using the pictures

- Ask the children to discuss how the structure of the seeds, or their container, enables a plant to disperse its seeds as shown in the captions.
- They could draw labelled diagrams and write explanations about how the structure of the seeds and their container (where there is one) enables the seeds to be dispersed: for example, the dandelion has tiny, light seeds, each with its own parachute-like structure, which enables the seeds to be blown on the wind for great distances; the strawberry has pleasant, sweet-tasting flesh around its seeds which encourages animals to eat them; the burdock's seeds have small barbs which make them cling to animals' fur, so that they can be carried by the animals for long distances; the poppy seed head dries out, as do the seeds, which escape from the small hole around the top of the seed head when the plant is shaken by the wind or by passing animals; the meadow cranesbill's pods pop open; sea rocket seeds are full of air, which helps them to float.

Activities

- Ask the children to find examples of other seeds and seed heads with similar structures to each of the ones in the pictures and to find out if these are dispersed in the same ways. They could make a display about the dispersal of seeds.
- To add interest, allocate different plants to different groups and ask them to find out as much as they can about their seeds. They could add these to the display or compile a database about seeds.

Higher-achieving children
Ask them to make 'fact cards' about a plant of their choice, which includes a photograph, information about its size, where it grows, its seeds and how they are dispersed.

Lower-achieving children
Draw their attention to the features of the seeds which enable them to be dispersed.

Plenary session

- Draw out that plants disperse their seeds in different ways so that as many as possible can survive.

Vocabulary

dispersal, disperse, perennial, reproduce, seed

Photo Book reference Page 13

10 Germination

QCA links
This develops the children's understanding about how plants reproduce and about the conditions needed for seeds to germinate. (QCA Unit 5B)

Background information

During the winter, seeds lie dormant in the ground. They need warmth and moisture (but not light) in order to germinate. Once they have grown a shoot they also need light, in order to produce their own food by means of photosynthesis. Seeds contain food which is used to produce the growing root and the first shoot.

Starting point

- The children should first have completed topics 8 and 9. Ask: *At what time of the year do most plants produce their seeds, and why?* The seeds are produced as the flowers die at the end of the summer. Discuss what happens to the seeds; some fall to the ground near the parent plant and some are dispersed. *Do they begin to grow immediately?* Draw out that seeds lie dormant over the winter. Introduce the term *germinate* for the process in which seeds begin to grow roots and shoots and ask the children what they know about the time of the year when seeds germinate.

Using the pictures

- Ask the children to describe what is happening in each picture. They could draw labelled diagrams and write explanations about how the seed has developed and how the plant has changed.
- *Do they think that the first root and shoot grew at the same time or that one grew before the other? Is this the same for all seeds?* The children could make notes about how they can find out.
- *What conditions do the children think a seed needs in order to germinate?* Ask them to write their ideas.

Activities

- Provide a large seed, such as a broad bean, cut in half to reveal the *embryo*:

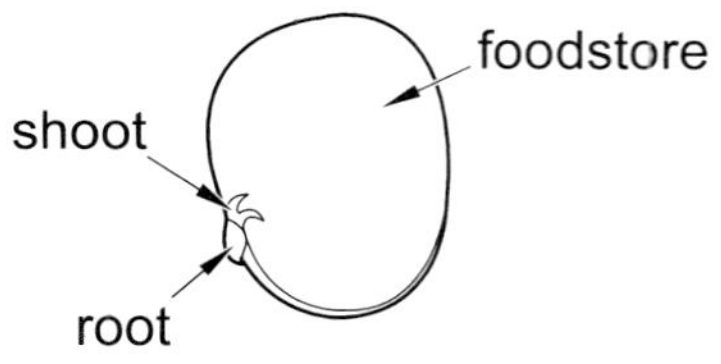

  Explain that the embryo does not change during the cold winter months and that germination is the development of the embryo into a plant.
- Ask some of the children to plan and carry out an investigation to find out what a seed needs in order to germinate. (Different groups could investigate different factors: light, water, warmth, soil). Others could plan and carry out an investigation to find out about the order in which the first root and shoot grow (or whether more than one root and shoot grow at the same time) in different plants: for example, broad bean, marigold, cress, mustard, tomato. So that the emerging roots and shoots are visible, plant the seeds close to the edges of clear containers. They will be easier to see if the containers are filled with cotton wool or paper towels.

Higher-achieving children
Ask them to use secondary sources to find out why seeds will germinate without the help of light but plants will not grow well without light. Draw out that plants use the green pigment, chlorophyll, to help them produce their own food from carbon dioxide gas in the air but that they can only do so where there is light.

Lower-achieving children
Help them to draw conclusions from the results of their investigations. Ask: *Did seeds germinate without water? Did they germinate without light? Did they germinate in the fridge? What did they need to help them to germinate?*

Plenary session

- Draw out that seeds need water and warmth, but not light, in order to germinate.

Vocabulary

dormant, embryo, germinate, germination, root, seed, shoot

11 Stamens of a lily

Photo Book reference Page 14

QCA links

This develops the children's understanding about how plants reproduce. They learn that plants produce flowers which have male and female organs and that seeds are formed when pollen from the male organ fertilises the ovum (female). (QCA Unit 5B)

Background information

The lily is a simple flower (its petals are arranged in a circle with the reproductive parts at its centre). Some flowers have either male or female parts; others have both. Sometimes an entire plant has male only, or female only, flowers. The lily in the photograph has both male and female parts. The orange stamens (male parts) can be seen clearly at the tips of the long green filaments, as can the green stigma at the tip of the thicker green style, which leads down to the ovary.

Starting point

■ Ask the children what they know about reproduction in plants. They will know from topics 7–9 that plants produce seeds from their flowers, that they disperse their seeds in different ways and about the conditions which help seeds to germinate. *Do they think there are male and female plants as there are male and female animals?*

Using the picture

❑ Ask the children to describe the flower in detail. (They do not need to be told the names of the different parts, yet.) Ask: *How many petals does it have, how are they arranged and what are they like?* They could comment on their texture, shape, colour and pattern.

❑ Ask them to describe the centre of the flower. *How many pieces protrude from it? What colour are they and what is on the tips of them? Are they all the same?* Draw attention to the one which is different from the others. *How is it different?* (Its tip is green and shiny and has a shape resembling three small balls fixed together, whereas the others have long orange, powdery tips.) *What is the orange powder?* (Pollen).

❑ Explain that the orange parts covered with pollen are stamens (male parts) and that the green tubular piece with the rounded end is the stigma (female part). See Activities.

Activities

- Tell the children that, in the summer, pollen from flowers is blown about by the wind. *Do any of them suffer from hay fever? What are the symptoms and when do they notice them the most?* (Usually when they are outdoors in gardens or fields in the summer.) Explain that hay fever can be triggered by pollen in the air and that, to help hay fever sufferers to know when to stay indoors, weather forecasts often include a pollen count. Provide newspapers or access to the Internet and ask the children to find out the pollen count for the current day. They could record it each day, over a period of time, and note on which the days any hay fever sufferers in the class have symptoms.
- Use ***Photocopiable sheet 2*** to explain how the plant is pollinated. Explain that pollen from the male part has to enter the stigma and reach the ovum; a seed grows from the fertilised ovum. The ovum cannot grow into a seed unless it is fertilised.

Higher-achieving children

Ask them to use secondary sources to find out about the stamens and stigmas of other types of flowers.

Lower-achieving children

Help them to relate the diagram in ***Photocopiable sheet 2*** to the lily in the picture.

Plenary session

◆ Draw out that a seed cannot grow unless the ovum in the female part of a flower is pollinated by pollen from a male part.

Vocabulary

female, fertilise, filament, flower, lily, male, ovary, ovum (plural ova), petal, pollen, pollinate, reproduce, stamen, stigma, style

12 Pollination

Photo Book reference Page 15

QCA links
This develops the children's understanding about how plants are pollinated by insects. (QCA Unit 5B)

Background information

Some flowers are wind-pollinated, but insects such as bees, wasps and butterflies help to pollinate others. The insects visit flowers to feed and, as they do so, the hairs on their bodies and legs pick up pollen from the stamens. The pollen brushes off on to the stigmas of flowers and makes its way down the style to the ovary, where it fertilises the ovules in the ovum. Each ovule can grow into a seed. The colours, patterns and scents of flowers attract different insects.

Starting point

- The children should first have completed topics 7–11. They could draw a concept map about plant reproduction. Ask them to identify anything they do not yet know: for example, how the pollen gets into the style and what happens inside the ovary.

Using the picture

- Ask the children what they think the bee is doing on the flower: bees feed on nectar from flowers and they carry nectar and pollen back to their nest or hive to feed their young (larvae).
- Draw attention to the insect's body and legs: they are covered with small hairs. Ask: *What happens as the insect moves about on the flower and brushes against the stamens?* Explain that some of the pollen which clings to the hairs brushes off on to the stigma and that this helps to pollinate the flower.

Activities

- The children could observe flowering plants outdoors and notice which ones are visited the most by insects. Ask them to choose a plant, or several plants of the same type, and record the number of insects which land on them. (They could make a tally chart.) This is best carried out on a warm sunny day. Different groups could observe different types of plant. Encourage them to think about how to make the test fair (by using the same number of each type of plant and observing them for the same length of time). Ask them to write a report about their investigation.
- They could try using a cotton bud to pollinate the flowers of plants growing in the classroom.

> ⚠ The children should be aware that some insects (bees and wasps) sting, but that if they sit still and do not try to touch the plants they are unlikely to be stung. Children who are known to suffer serious reactions to bee or wasp stings should not take part.

Higher-achieving children
Ask them to plan a small garden to attract specific insects: for example, butterflies.

Lower-achieving children
Discuss why the 'insect count' should be timed and why the same number of flowers of each type should be observed. It might help if they are given a format on which to write their reports: for example, a chart with the headings: *Our question, What we did, How we made the test fair, Our results, What we learned…*

Plenary session

- Draw out that some plants are pollinated by insects and that they attract insects by their shape, pattern, colour and scent.

Vocabulary

female, fertilise, filament, flower, lily, male, nectar, ovary, ovule, ovum (plural ova), petal, pollen, pollinate, reproduce, stamen, stigma, style

13 Inside an egg

Photo Book reference Page 16

QCA links

In this activity the children learn about animal reproduction: that adult animals have young and that, if they did not, the species would eventually die out. (QCA Unit 5B)

Background information

The hens' eggs which we eat have not been fertilised by sperm from a male bird and so will not have chicks developing inside them. The yolk is produced in the hen's liver and passes down the oviduct, surrounded by a membrane. As it does so, layers of albumen form around it. Later in the journey down the oviduct a tough membrane, and then the shell, develop around the albumen. Strands named *chalazae* anchor the yolk within the shell. The process takes a day. A fertilised egg is laid the day after mating. At this point the embryo looks like a small dark speck. Blood vessels and then a tube (the yolk stalk) develop in the yolk. A sac named the *allantois* develops; this acts as a lung as well as storing waste products from the developing chick. The eggshell is porous; oxygen and waste gases can pass through it. After three weeks a small air pocket develops at the wider end of the egg to enable the developing chick to breathe. The chick's beak has a hook which helps it to break out of the shell; this disappears after a few days.

Starting points

- Ask the children what they know about plant and animal reproduction. *In what ways are they similar?* In some plants and animals ova in the female have to be fertilised by material from the male (pollen in plants and sperm in animals) so that a new life can begin.
- The children could list animals which lay eggs and animals which give birth to live young. They could also make a list for animals about which they are not sure: for example, some snakes lay eggs, but in others the eggs hatch within the body of the female.

Using the pictures

- Ask the children to work with a partner to describe the process represented in the photographs. Ask them to draw a flow chart in order to present the process of the development of the chick. *Can they turn the flow chart into a cycle? What will happen to the chick?* (It will grow into an adult bird.) *How can they link this to the stage presented at the start?* The adult will mate with a bird of the same species, but the opposite sex, and a fertilised egg will result.
- The children could draw and label the life cycle of the bird, add annotations in boxes (this could be linked with work in literacy lessons).

Activities

- Provide an egg for each group to examine, using magnifying glasses. Crack the shell and empty the contents into a white saucer. *Can the children identify the yolk, albumen, chalazae and membranes?* They could make a labelled diagram of an egg.

> ⚠ The children should avoid touching their faces and should wash their hands after handling raw eggs.

- The children could use secondary sources to find out about the life cycles of other animals and present them as cyclical charts: for example, adder, bee, cat, crab, moth, sheep, snail, worm.

Higher-achieving children

Ask them to use secondary sources to find out more about the structure of a bird's egg.

Lower-achieving children

Provide a word bank to help them to label their diagrams.

Plenary session

- Draw out that all animals reproduce and that some lay eggs, whereas others give birth to live young.

Vocabulary

albumen, allantois, chalazae (singular chalaza), chick, female, fertilise, male, mate, membrane, ovary, ovum (plural ova), pollen, reproduce, sac, shell, species, sperm, yolk

Photo Book reference Page 17

14 A rat in the uterus

QCA links
In this activity the children learn about animal reproduction: that adult animals have young and that, if they did not, the species would eventually die out. (QCA Unit 5B)

Background information

Rats can breed at the age of three to four months and can produce up to seven litters, of 6 to 22 young, per year. The size of the litter depends on the size of the mother but the average is between 6 and 9. The second or third litter is the largest and, after the tenth, the litter size decreases sharply. The gestation period averages 21 days.

Starting point

- Ask the children what they know about reproduction in mammals. They could present their ideas in the form of a concept map. Ask: *How is mammal reproduction different from that of birds? How is it similar?*
- Draw out that in most animals reproduction involves the mating and the combination of sperm from a male animal and an ovum (egg) from a female. In most mammals the egg develops inside the female, and live young are born. In birds (and many other animals) the egg develops outside the body of the female.

Using the picture

- Ask the children to work with a partner to describe what they can see in the photograph. *Can they make out any parts of the foetus's body?* They should notice that the foetus develops inside a fluid-filled sac. Point out the membrane which surrounds it and protects the unborn rat.
- Ask them to use secondary sources to help them to draw a flow chart to present the life cycle of a rat.

Activities

- Ask the children what they know about the life cycle and reproductive methods of other mammals: for example, their pets. They could bring in photographs of any young produced by their pets, along with information about both parents. Discuss the characteristics the young have inherited from each parent. Note that animals breed with others of the same species.
- The children could use secondary sources to find pictures of the developing foetuses of other mammals. Ask them to compare this process with the development of other kinds of animals: birds, fish, insects, molluscs, reptiles. Draw out that mammals give birth to live young.
- The children could use secondary sources to find out about the life cycles of other mammals and present them as cyclical charts: for example, cat, chimpanzee, cow, human, mouse, rabbit, sheep.

Higher-achieving children
They could use secondary sources to help them to add a timescale to their flow charts of the life cycle of the rat.

Lower-achieving children
Provide a word bank to help them to label their diagrams.

Plenary session

- Draw out that rats, like humans, are mammals and that all mammals give birth to live young.

Vocabulary

foetus, mammal, mate, ovum, reproduction, rodent, species, sperm, uterus, womb

Photo Book reference Page 18

15 Liquids and gases

QCA links
This is about the differences between solids, liquids and gases. (QCA Unit 5C)

Background information

In theory, all materials can exist in any of the three states: solid, liquid or gas. Generally, they exist in one state: for example, we think of metals as solids, but they can be heated to high temperatures and changed to liquids. Water is the most commonly observed material in all three states, because its freezing point is not particularly low (0°C) and its boiling point is not particularly high (100°C). Heating changes a material from solid to liquid or liquid to gas by giving energy to the particles (atoms and molecules) of which it is composed. These particles are in a constant state of motion: they move about more quickly in a liquid than in a solid and more quickly in a gas than in a liquid. The bonds between the particles of a solid are more rigid and fixed than those between the particles of a liquid or gas, so a solid keeps its shape when turned out of a container but a liquid flows downwards and a gas escapes in all directions.

Starting points

- Ask the children to name some solid materials. *Are they always solid?* Discuss what has to be done to solid materials to make them liquid. *Can they name some liquids? What has to be done to them to make them solid?*
- Ask them what they know about gases. *Can they name any? Which gas do they need in order to stay alive?* (Oxygen, from the air, which is a mixture of gases.).

Using the picture

- ❑ Ask the children to list all the solids and liquids they can see in the picture:

Solids	Liquids
sand	water
rock	petrol
rubber	bubble mixture
ice cream	melted ice cream
metal	lemonade
wool	
cotton	
plastic	
chocolate	

- ❑ Ask: *Where are there gases?* They could list the places, even if they do not know the names of the gases: for example, in the air, aqualung, balloon, fizzy drinks, inflatable dinghy and tyres.
- ❑ Ask the children if they know what Americans call petrol (gas) and what is odd about this. (Petrol is a liquid.)

Activities

- The children could make a display chart of solids, liquids and gases, using pictures cut from magazines and newspapers.
- Ask them to list liquids and gases which are useful to us, in addition to air and water: for example, for heating, cooking, cleaning, and so on.

Higher-achieving children
They could use secondary sources to find the names of some of the gases in the air and in everyday objects, such as light bulbs, fairground balloons, camping gas cylinders, fizzy drinks, exhaust from cars. Ask them to list any which are harmful.

Lower-achieving children
Provide a word bank to help them to label their diagrams.

Plenary session

- ◆ Draw out that material can exist as solids, liquids or gases.

Vocabulary

air, gas, helium, liquid, oxygen, solid, state

Gases around us

Photo Book reference Page 19

16 Natural sponge

QCA links
Here the children learn that sponges are solid materials with air in the gaps between particles of material. (QCA Unit 5C)

Background information

Sponges are primitive aquatic animals. Early naturalists defined them as plants because of their branching form and because they did not seem to be able to move independently. Many modern zoologists regard them as having a separate classification from the rest of the animal kingdom. A sponge is a mass of cells and fibres, with a system of tubes leading to holes of different sizes, through a tough skin. Commercial sponges are produced by cleaning out millions of living cells, leaving a soft and elastic skeleton. Sponges can regenerate lost parts, so they can be cultivated from fragments. Synthetic sponges are made from plastics into which a gas is forced.

Starting points

- Ask the meaning of *absorbent*. Ask the children to contribute to lists of objects made from absorbent and non-absorbent materials, for example:

Absorbent	Non-absorbent
towel	plastic bag
kitchen roll	rubber boots
tissue	waxed jacket
cotton wool ball	metal pipe
sponge	glass tumbler
dishcloth	china cup
chamois leather	ceramic mug

- Draw on their previous learning about absorbent materials: ask them if they know how these materials differ from those which are not absorbent. Ask: *Why do some materials soak up water or allow it to pass through them, whereas others do not?* Draw out that water passes into or through gaps or holes; if there are no gaps or holes in material, water cannot pass into or through it.

Using the picture

- ❑ What does the caption tell us about this type of sponge? Ask the children what *natural* means. Natural materials can come from plants, animals or the Earth. *Where do they think this natural material comes from?* Tell them that sponges are animals (see Background information and **Resources**).
- ❑ Ask the children to describe the sponge. *Is it absorbent? How can they tell?*

Activities

- The children could find out what happens when they put a sponge into water. *What happens when they squeeze the sponge?* Ask them what causes the bubbles and why there should be air in the sponge.
- Ask them how the sponge changes when it is immersed in water. *How can they find out how much water it soaks up?* ***Photocopiable sheet 3*** suggests a way of measuring and recording the amount of water soaked up by a sponge.

Higher-achieving children
The children could work out the relationship between the weight of the sponge and the weight of the water it absorbs, and then compare this with the amount of water absorbed by artificial sponges.

Lower-achieving children
Before they begin the investigation, provide practice in using weighing scales and taking accurate readings.

Plenary session

- ◆ Draw out that a sponge contains air in its many cavities and that when a sponge is placed in water the air escapes from these cavities and is replaced with water, which is heavier than air.

Vocabulary

absorbent, air, bubble, cavity, gas, natural, non-absorbent, sponge

17 Air spaces

Photo Book reference Page 20

QCA links

Here the children learn that many materials have air trapped inside them or gaps between their particles, into which, air passes. They have opportunities to make careful observations of materials and to explain these using scientific knowledge and understanding. (QCA Unit 5C)

Background information

Pumice is formed when lava (molten rock which has emerged from the earth's crust) cools very quickly, allowing no time for crystals to form. As the rock solidifies, gases dissolved in it are released quickly, forming a froth which immediately solidifies. Lava which cools more slowly, or under greater pressure, forms obsidian – much denser, glass-like rock. In some cheeses, bubbles of carbon dioxide are produced by bacteria during ripening. Sponge cake mixture is beaten to trap air. It also requires an ingredient which foams and retains the air bubbles, such as baking powder or egg white, or both.

Starting point

- Before the children look at the pictures, ask them if they can think of any similarities between pumice, Jarlsberg cheese and sponge cake. *Do they know what pumice is?* Ask them to discuss this question in their groups for a few minutes and to make notes. Then invite feedback before they look at the pictures.

Using the pictures

- Once the children have seen the pictures, can they spot a similarity between the materials? (They contain air holes.) Ask them if they can explain how the air holes got into the materials. They could write their ideas and then use secondary sources to find out.
- Draw out that these solid materials were once liquid: pumice solidified as it cooled, trapping bubbles of gas. Cheese is made from a liquid (milk) and a new material is formed with bubbles of gas in it. Sponge cake mixture is liquid with some solid materials dissolved and others mixed with it. A new solid material and a gas are formed when it is heated.

Activities

- Ask the children if they think a piece of Jarlsberg cheese will weigh the same as a piece of another type of cheese of the same size. *Will the sponge cake weigh the same as a piece of fruitcake of the same size? Do the air holes make pumice lighter for its size than other rocks?*
- Ask them to write and explain their predictions before using a weighing balance to find out. Help them to explain their results.
- The children could compare sponge cake mixture uncooked and after baking. Discuss what they will observe. *What do they think will change?* They should weigh it, measure its height in the baking tin and make observations about its texture and colour.
- *Do the children think air makes materials lighter? Does air have any weight?* Ask them if they think an empty balloon will weigh the same after it has been filled with air. ***Photocopiable sheet 4*** suggests a method of finding out.

Higher-achieving children

Challenge them to find out if a fizzy drink weighs the same as the same drink without fizz, and to explain their predictions and results.

Lower-achieving children

Use this as an opportunity to develop skills in accurate measurement of height and weight.

Plenary session

- Explain that the holes in materials which were once liquid are formed when bubbles of gas are trapped in the liquid and that air is lighter than most materials, including water. This is why materials with many air holes are lighter *for their size* than other similar materials without air holes. Note that air has weight; adding air does not make something lighter. A balloon becomes heavier when it is inflated, but it also becomes a lot bigger; air makes it lighter *for its size.*

Vocabulary

air, bubble, carbon dioxide, foam, gas, lava, molten, pumice, rock, sponge, volcanic, volcano

Photo Book reference Page 21

18 Useful gases

QCA links
Here the children learn that there are many gases and that several of these are important to us. (QCA Unit 5C)

Background information

An aqualung contains compressed air. Gases can be forced into small spaces under pressure. This makes it possible for more air than would naturally fill the cylinder to be stored in it. Our bodies need oxygen from the air we breathe in. Air contains many other gases which we do not need; we exhale them, along with carbon dioxide produced in the lungs during respiration. The patient is being given pure oxygen, for more efficient respiration. Camping gas lamps and stoves burn butane and propane gas from small cylinders, where it is stored in liquid form. It is not difficult to reduce the temperatures of the gases enough to liquefy them for storage in pressurised containers. They vaporise once released.

Starting point

■ Before the children look at the pictures, ask them to contribute to a list of the uses of gases at home, in factories, in transport, hospitals, and so on: for example, energy to provide light and heat, for making fizzy drinks, filling tyres, dinghies, balloons and other inflatables, for anaesthetics, in air brakes.

Using the pictures

❑ Discuss the purposes for which the gases are being used (breathing and light). *Can they name the gases?* (Compressed air, oxygen, propane and butane.)

❑ Ask: *What would happen to each gas if the container were opened? Would the same happen to a liquid (such as water) or a solid (such as sand) in similar containers?* Draw out the differences between the ways in which solids, liquids and gases behave.

Activities

- The children could sort a collection of utensils and containers on to a Venn diagram made from hoops, according to what could be kept in them: solids, liquids or gases:

What can containers hold?

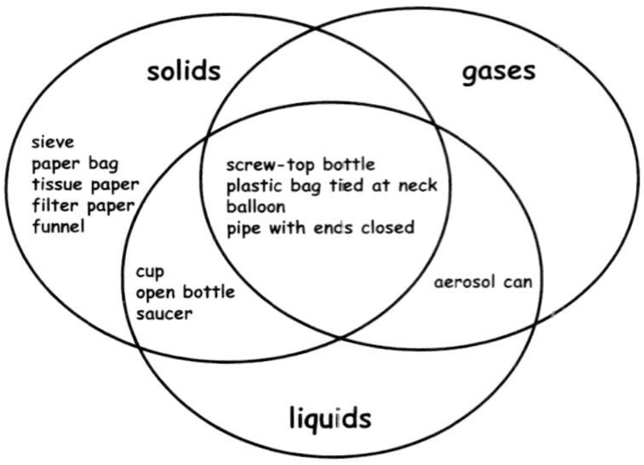

Higher-achieving children
Challenge them to use secondary sources to find out the names and characteristics of different gases and what makes them suitable for different purposes: for example, in electric light bulbs or fluorescent tubes, cooling a refrigerator, making drinks fizz.

Lower-achieving children
After they have sorted the containers for different materials, ask them which form of material can be placed on a completely flat surface, which needs a container with sides and which also needs a lid, and why.

Plenary session

◆ Draw out that gases can move in all directions; liquids flow downwards, even down a very gentle slope, and escape through very small holes; solids fall downwards if not supported, but do not flow, and can be held in containers with holes smaller than the pieces of solid.

Vocabulary

air, butane, gas, lava, oxygen, propane

Gases around us

Photo Book reference Page 22

QCA links
The children learn that there are many gases, that several are important to us and that liquids other than water evaporate to form gases which flow easily from place to place. They make observations and explain them using scientific knowledge and understanding. (QCA Unit 5C)

19 Spray

Background information

Sprays use air to force liquid up a tube and out of a nozzle. Aerosols have gases in the canister, along with the liquid they force out. Scent evaporates quickly.

Starting point

- Discuss the similarities between a water pistol, a washing-up liquid bottle and a spray polish. (They all use gases to force a liquid through a small opening.)

Using the picture

- Ask the children what is happening in the picture. Ask: *Is the scent in the bottle a solid, a liquid or a gas?* Ask them what they know about the ways in which materials in each state behave (see **Resources**).
- Ask: *If the top were taken off a bottle of scent, would the liquid escape?* Remind the children about how liquids behave. *How can we tell that the scent is coming through the nozzle?* (We can see droplets in the air.) *Liquids do not move upwards, so what is happening here?*
- Discuss what will happen to the scent droplets which land on the woman's skin. (They will evaporate.) *Will anything be left on her skin?* Draw out that something must remain so that the scent stays (some material will evaporate more slowly).
- Discuss how the woman can smell the scent. Ask: *With which organ does she smell it? How does the scent reach her nose?* Remind them that evaporation is what happens when a liquid becomes a gas, and how gases behave when they escape into the air. Gases enter our noses when we breathe in. Explain that cells inside the nasal passages receive smells and send messages to the brain, which can identify different smells.

Activities

- Provide some sprays (not aerosols) for the children to examine and to draw annotated diagrams to explain how they think they work.

> Warn the children that aerosols can be dangerous. They explode if heated. The gases in them can be harmful, and even lethal, if breathed in.

- The children could use the indexes of information books and CDs, and carry out Internet searches to check their observations about how sprays work. Point out that they are different from aerosols and show the children some aerosol containers for comparison. Draw out that air can be used to force liquids along tubes and through narrow openings. ***Photocopiable sheet 5*** suggests a method of testing perfumes for the strongest smell (the one which can be detected from the greatest distance). Only one group at a time can work on this in the same room. The flow chart in ***Photocopiable sheet 6*** supports the writing of an explanation of how a spray works.

Higher-achieving children
They could research and write an explanation (with a labelled diagram) of how we smell things.

Lower-achieving children
Help them to write their explanation of how the spray works with the aid of a flow chart.

Plenary session

- Draw out the ways in which gases, including air, are useful. Also focus on the role played by evaporation in the sense of smell.

Vocabulary

aerosol, air, evaporate, force, gas, liquid, nozzle, perfume, scent, spray

20 Solid, liquid or gas?

Photo Book reference Page 23

QCA links

Here the children learn that gases are different from solids and liquids in terms of how they do not maintain their shape and volume and that gases flow more easily than liquids and in all directions. They identify and describe differences in properties of solids, liquids and gases. (QCA Unit 5C)

Background information

The balloon on the left of the photograph contains ice; the balloon was attached to a cold water tap, filled with water and then frozen overnight. The balloon in the centre contains water and the one on the right contains air.

Use a balloon pump. Never let the children inflate a balloon by blowing into it. This can cause choking if they breathe in by mistake. Wear gloves when removing a balloon from the freezer. Do not let the children handle it immediately.

Starting point

- Ask the children what they would do to find out whether a sealed container is filled with a solid, a liquid or a gas. *Which senses would they use? What clues would they find?*

Using the picture

- Ask the children what they think is inside the balloons. Draw attention to the effects of the boys' fingers on the balloons. Ask: *Are the balloons filled with a solid, a liquid or a gas?* Tell the children what is in each balloon and ask them about the differences between the materials. Ask: *What would happen to the materials if each balloon were pierced with a pin? Would they keep their shape, remain on the table or flow away (and, if so, in which directions)?*

Activities

- Ask the children to plan a series of tests to compare the contents of the three balloons: one filled with ice, one with water and one with air. *What will they observe and measure? How would they make the comparison fair?*

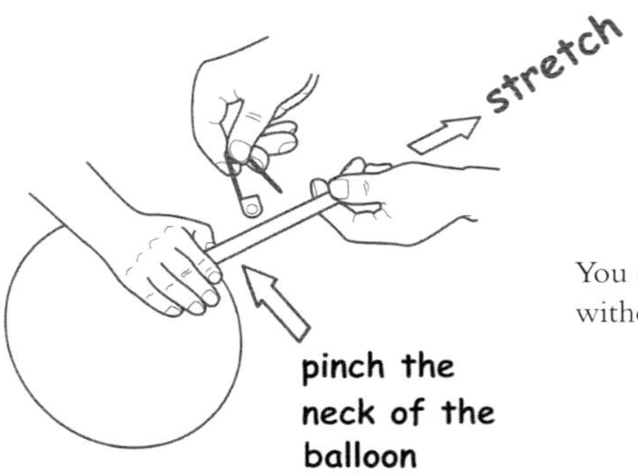

You can pierce a balloon without a bang!

- They should consider the sizes of the balloons, their weight when filled, whether or not they will float in water, what they feel like and what will happen when they are pierced with a pin. They could predict and then test the balloons and record their findings on a table. Ask them to use their findings to help them to write a comparison of solids, liquids and gases.

Higher-achieving children

They could find out more about the way in which ice floats. How much is above and how much is beneath the surface of the water? Ask them to compare this with other materials which float: for example, balsa wood, polystyrene, oak (or pine, mahogany or another wood).

Lower-achieving children

Provide a format on which they can plan their investigation or headings to help them: for example, *What we are comparing, What we shall do, What we shall observe and measure, Equipment we shall need.*

Plenary session

- Draw out the characteristics of solids, liquids and gases: in particular, that solid materials keep their shape, whereas liquids flow downwards when they escape from a container and gases flow away more easily and in all directions.

Vocabulary

air, escape, flow, gas, heavier, lighter, liquid

21 Water (1)

Photo Book reference Page 24

QCA links
This consolidates the children's understanding of the characteristics of materials in different states, focusing here on water. They learn that water collects in streams and rivers and eventually finds its way to the sea. (QCA Unit 5D)

Background information

There is no agreed definition of a stream and a river, but any flowing body of water which is narrower than the small river shown in the bottom photograph (the river Tees near its source in Cumbria) is usually referred to as a stream. A spring is the source of a river or stream.

Starting points

- This could be linked with work in geography on water. Ask the children what happens to water which falls as rain. Ask: *Where does it land? What happens to it after that?* Discuss how long the water remains and whether this differs from place to place: for example, on different types of roofs, tarmac, concrete, paved surfaces, rock and soil.
- Draw out that some water evaporates but some soaks into the ground. *Where does it go?* Some flows along the ground. *To where?*

Using the pictures

- Ask the children to list the similarities and differences between the bodies of water in the photographs. They could think about width, depth and speed of flow.
- Ask them to list the photographs in order and to explain this order. They could draw a flow chart of the growth of a river from source to the sea.
- *Where does the water in the sea come from?* Draw out that it comes from many rivers, and ask the children why they think rivers flow towards the sea. Remind them about their previous learning about the characteristics of liquids and that water is a liquid. On a physical map of Britain they could find the altitude of the land in different places and notice where the rivers rise. *Why do some rivers with sources in the same area flow in one direction and others flow in other directions?* Ask them to find the watershed and explain why it is in that location. (The watershed runs along the highest ground.) Draw out that water always flows downhill. Point out that there are some places where the land is below sea level and that rivers flow into these places and form lakes. Streams and rivers flow out of lakes towards the sea.

Activities

- The children could carry out a survey about what happens to rainwater which falls around the school. Ask: *Where does it dry up the most quickly? Where does it collect in puddles? Where does it dry up the most slowly?*
- Ask them to make observations about these places and to explain their results. *Are the places low-lying, higher than the rest of the ground or on a slope? Do they have much sunlight? What material is on the surface of the ground?*
- ***Photocopiable sheet 7*** suggests a method of investigating different types of soil to find out how well water soaks into them. The children could use the results to help them to explain their findings in their survey.

Higher-achieving children
Ask: *What could a gardener or farmer do to help water to soak into the clay soil of a waterlogged field?* They could test their ideas using small amounts of clay soil and write a report about their findings.

Lower-achieving children
Provide a model to help them to explain watersheds: they could sprinkle water on to a long, clay 'hill' in a bowl and, on a diagram, mark the directions in which the water flows. Encourage them to experiment by changing the shape of the hill.

Plenary session

- Draw out that water flows downwards and that rivers flow from high ground towards the sea because the sea is usually lower than the land.

Vocabulary

flow, river, spring, stream, water, watershed

22 Water (2)

Photo Book reference Page 25

QCA links
Here the children learn that water evaporates from oceans, seas and lakes and eventually falls as rain. They turn ideas into a form which can be investigated, make a prediction and decide what evidence to collect. (QCA Unit 5D)

Background information

The photographs represent parts of the water cycle. Clouds consist of condensed water vapour suspended in the atmosphere. Three main types of cloud are *cirrus* (high, wispy strands), *cumulus* (heaps of rounded lumpy cloud) and *stratus* (horizontal sheets). These terms are combined to name different forms of cloud: for example, *cumulonimbus*, as in the photograph. *Precipitation* refers to any moisture which condenses from the air: for example, *dew, frost, hail, rain, rime, snow*. Hundreds of droplets of water coalesce to form a raindrop; it absorbs other droplets as it falls. *Frost* is made up of ice crystals formed by frozen *dew*. *Rime* is a crust of ice crystals formed on anything touched by freezing fog. Precipitation eventually reaches rivers or is intercepted and used by humans (or drunk by other animals). Water which passes through any system – animal or industrial – returns to the water cycle.

Starting points

- The children should first have completed topic 21. Play a recording of a weather forecast and ask them to notice the technical subject words (even the simple ones, such as *rain, sunshine* and *snow*). Invite feedback and compile a list of these words.
- Ask the children to use the words in a concept map about water: encourage them to think about rainfall. Ask: *Where does rain come from? What causes it? What happens to all the water which lands on the earth, including water used by people, plants and animals?*

Using the pictures

- Ask the children what they know about the water cycle and draw on the concept maps they made (see Starting points). Explain *dew, frost, precipitation, rime.*
- In pairs, they could discuss how each photograph fits into the water cycle.
- Ask them to write a sentence to link each photograph to the others and to the water cycle. Draw out that all the water which is in the Earth and its atmosphere has been here since the beginning of time and has been recycled over and over again. The total amount has changed very little.

Activities

- Ask: *If water from the sea evaporates, forms droplets of water vapour which become clouds then condenses and falls as rain, will that rain be salty?* Ask the children to plan an investigation to find the answer. ***Photocopiable sheet 8*** suggests a method of turning the question, 'Will rain, from water evaporated from the sea, be salty?' into a form which can be investigated. Introduce the idea of using a model to investigate a question which would otherwise be difficult to investigate. Ask the children what they could use to represent seawater, how they could evaporate it and how they could collect the evaporated water when it condenses. Remind them that they need not boil the water in a pan or kettle –the heat of the Sun can heat it enough. *How can they magnify the heat of the Sun?*

Higher-achieving children
They could find out and explain how rain, for example acid rain, becomes polluted.

Lower-achieving children
Remind them of their previous work on changes of state. Ask: *Where does water go when it evaporates? How can it be collected before it escapes into the air?*

Plenary session

- Draw out that water continually changes from liquid to gas and sometimes to solid, and back again, and that when it evaporates any solid material in it is left behind. Also point out that rain can become polluted as it falls, when gases in the air dissolve in it.

Vocabulary

cloud, condensation, condense, dew, evaporate, evaporation, frost, hail, precipitation, rain, rime, snow, water cycle, water vapour

23 Changing

Photo Book reference Page 26

QCA links
The children learn that evaporation is when a liquid becomes a gas, condensation is when gas becomes a liquid, freezing is when a liquid becomes a solid, melting is when a solid becomes a liquid, condensation is the reverse of evaporation and that melting is the reverse of freezing. (QCA Unit 5D)

Background information

The pictures represent changes of state; some are in the process of taking place and others have already taken place. For example, the ice cream is melting, the water in the kettle is boiling and evaporating, water is condensing on the cold containers; the water has evaporated from the soil and the ice lollies have frozen.

Starting points

- Introduce the topic with ***Photocopiable sheet 9***, which provides a format for a game for groups of 4 to 6. They choose letters at random (for example, by opening a book and taking the first letter on the page) and, within a limited time (say 30 seconds), try to think of a solid, liquid and gas beginning with that letter. They score 6 points for an answer which no one else in the group has, 5 for an answer shared with one other person, 4 for an answer shared with 2 others, and so on.
- Invite volunteers to read out an answer for a solid; then ask if anyone also wrote the liquid or gas form of that solid anywhere on their answer chart.

Using the pictures

- Ask the children to write captions on stickers to show which change of state is taking place, or has taken place, in each photograph. Invite feedback.
- Ask them to work with a partner to list pairs of photos which show opposites and to explain why they have paired them: for example, ice cream and ice lollies because the ice cream is a solid becoming a liquid (melting) and the ice lollies are made from a liquid becoming a solid (freezing); the boiling kettle and the cold drinks because the water in the kettle is becoming a gas (evaporating) and a gas (steam) is becoming a liquid (water) on the sides of the containers (condensing). They could also pair the dried-up ground (evaporation) with the cold containers (condensation).
- Draw out that the photographs show *changes of state*. Introduce or revise this term.

Activities

- Make ice cream with the children, using a freezer or an ice cream maker. Begin as early as possible in the day. Ask them to predict what will happen as the temperature of the mixture is reduced. Using a sterilised thermometer, which is kept for use with food, they could check the temperature of the mixture throughout the day and record the temperatures on a line graph. *At what temperature does the mixture begin to become solid?* ***Photocopiable sheet 9*** provides a format for a game about solids, liquids and gases (see Starting points).

Higher-achieving children
They could use a datalogger to record the temperature changes of melting ice cream. Use ice cream which is not going to be eaten.

Lower-achieving children
Remind them of their previous work on changes of state. Help them to check the temperature of the air in a fridge and a freezer and to explain why frozen foods are kept in the freezer and cold foods in the fridge.

Plenary session

- Draw out that freezing is the opposite of melting and evaporation is the opposite of condensation, and that these are all reversible changes of state.

Vocabulary

boil, change of state, condensation, condense, evaporate, evaporation, freeze, gas, liquid, melt, reversible, solid, steam, water vapour

24 Drying

Photo Book reference Page 27

QCA links
The children identify factors which affect evaporation and explain everyday examples of drying in terms of factors affecting evaporation. They turn questions about evaporation into a form which can be investigated; they plan and carry out a fair test and present their results in the form of a chart. (QCA Unit 5D)

Background information

The laundry dries more quickly in sunny, windy weather than in dull, cloudy, damp weather. Humid air already contains a lot of water and cannot absorb more. Warm air can absorb more moisture than cold air. Wind helps in the drying process by whipping water molecules into the air and moving the moist air away from the wet material, allowing dry air to move in and to absorb water. When the air is still, moisture does not move away so easily.

Starting points

- As a homework activity, the children could prepare for this topic by asking people they know how they dry their laundry, which method they prefer, and why.
- Invite feedback from the children and list different methods of drying laundry: for example, in a tumble dryer, on a clothes line, on a rotary line, on a clothes horse, on radiators. Ask: *Which method did people say they preferred, and why?* Discuss where the water from the damp laundry goes.

Using the pictures

- Discuss what happens to the laundry in these photographs. Ask the children what they mean by *drying*. Ask them how wet and dry laundry is different. Draw out that wet laundry has absorbed water but when it dries the water has gone. *Where does the water go from the laundry on the clothes line? Where does it go from the laundry in the tumble dryer?*

Activities

- Ask the children to plan an investigation to compare different methods of drying laundry. Point out that instead of using real laundry, tumble dryers and clothes lines, they can set up a fair test which uses models.
- Point out that it is difficult to tell when fabrics are completely dry, but that if the children use coloured absorbent sugar paper they can easily see when it is dry and that it is easy to keep the samples the same size and weight. ***Photocopiable sheet 10*** suggests some drying methods which could be used as models to represent real-life laundry drying and suggests a method of recording the investigation.

Higher-achieving children
They could predict, and then find out from manufacturers' leaflets, how the water from laundry in tumble dryers escapes. Ask them to find three different methods.

Lower-achieving children
Ask them how they will be able to tell when a piece of sugar paper is dry. *Which method do they expect to be the quickest and what do they expect to see happening if they are correct?*

Plenary session

- Ask the children to report the results of their investigations. Discuss any discrepancies in the results and ask the children how well the investigation answered the question.

Vocabulary

change of state, drying, evaporate, evaporation, gas, humid, liquid, moisture, water vapour

Photo Book reference Page 28

25 Melting ice

QCA links

The children have opportunities to make careful measurements of temperature and record them on graphs. They learn that ice melts (and water freezes) at 0°C. (QCA Unit 5D)

Background information

Ice begins to melt at 0°C and remains at this temperature throughout the process of melting. Once the ice has melted, the resulting water rises in temperature until it reaches the same temperature as its surroundings.

Starting point

- Ask the children what they know about ice. They could contribute to a concept map about ice.

Using the pictures

- Mask the graph and ask the children to look at the picture of the beaker of ice.
- Ask them to read the temperature on the thermometer and to write instructions for using a thermometer. *Can they predict what will happen to the temperature if the beaker of ice is left in a warm room?* Ask them to draw the shape of a line graph representing ice melting. Invite volunteers to present their ideas for the graph and to explain what it shows.
- Ask them to uncover the picture of the graph and to compare its shape with those of the rough graphs they drew. *What do they notice about the shape of the graph?* Ask: *What is happening where the graph has a straight horizontal line?* Discuss anything which surprised the children and explain that throughout the time it is melting, ice has a temperature of 0°C.

Activities

- The children could check the temperature of melting ice and draw their own graphs. Ask them to compare their graphs with those of others in the class and with the one in the picture.
- Ask: *What will happen to the temperature of water as it freezes?* Remind them that freezing is the reverse of melting. They could draw the shape of the graph they expect to see and then plan an investigation to find out, using a datalogger with a temperature sensor.
- They should find that the temperature of the water drops to 0°C and remains at that temperature for some time (as when the ice melted). *What do they think was happening during that time?*

Higher-achieving children

Ask: *What do people sometimes spread on icy footpaths to make the ice melt quickly? Do they think grit or salt can melt ice?* They could plan an investigation to find out if ice melts more quickly if salt is added to it and whether this affects the melting and freezing temperatures of water.

Lower-achieving children

Show them how to use a thermometer and provide practice in taking the temperatures of materials such as cold water and warm water before they begin the investigation.

Plenary session

- Ask the children to report the results of their investigations. Discuss any discrepancies and ask them what they have learned about the melting and freezing of water.

Vocabulary

Celsius (C), change of state, degrees, freeze, ice, liquid, melt, temperature, thermometer, water

26 Boiling water

Photo Book reference Page 29

QCA links
The children have opportunities to make careful measurements of temperature and record them on graphs. They learn that water boils and evaporates (and steam condenses) at 100°C. (QCA Unit 5D)

Background information

Pure water, at normal atmospheric pressure, boils and begins to evaporate at 100°C and remains at this temperature throughout the change of state. It cannot reach temperatures higher than 100°C at normal atmospheric pressure. Steam condenses at 100°C. Once the steam has condensed, the temperature of the resulting water reduces until it reaches the same temperature as its surroundings.

Starting point

- Ask the children what they know about steam. They could contribute to a concept map about steam.

Using the pictures

- Mask the graph and ask the children to look at the picture of the boiling kettle.
- Ask them to read the temperature on the thermometer. *Can they predict what will happen to the temperature if the kettle continues to boil and does not have an automatic cut-off switch?* Ask them to draw the shape of a line graph representing water being heated and then left to boil. Invite volunteers to present their ideas for the graph and to explain what it shows.
- Ask them to uncover the picture of the graph and to compare its shape with those of the rough graphs which they drew. *What do they notice about the shape of the graph?* Ask: *What is happening where the graph has a straight horizontal line?* Discuss anything which surprised the children and explain that throughout the time it is boiling, water has a temperature of 100°C. It cannot get any hotter than this because water at a temperature higher than 100°C would be steam and would have escaped from the kettle.

Activities

- The children could check what happens to the temperature of water as it is heated, using a datalogger with a temperature sensor, and print out graphs which show the changes in temperature. Tap water can be used but for greater accuracy use distilled water. Ask the children to compare their graphs with those of others in the class and with the one in the picture.

> ⚠ Place the kettle well away from the children and put a table in front of it to prevent accidents.

- Remind them that evaporation is the reverse of condensation and ask them what they would expect to see on a graph showing steam condensing. To check this they could use simulation software.

Higher-achieving children
Ask them to make observations on the behaviour of steam as it escapes from the kettle. Ask: *How does it affect the lid of the kettle?* Link this with the children's experience of pans boiling and their lids lifting. They could use secondary sources to find out about the ways in which the power of steam is used and how it was used in the past.

Lower-achieving children
They could first practise using a datalogger and printing line graphs. Discuss why a line graph is better than other forms of graph (such as a bar chart) for recording a changing temperature.

Plenary session

- Ask the children to report the results of their investigations. Discuss any discrepancies and ask them what they have learned about the evaporation and condensation of water.

Vocabulary

boil, Celsius (C), change of state, condense, degrees, evaporate, liquid, steam, temperature, thermometer, water

27 Changes

Photo Book reference Page 30

QCA links
The children learn that snow is a solid form of water – made from crystals of ice. They also learn that other materials, as well as water, change state and about the ways in which people find these changes useful. (QCA Unit 5D)

Background information

High clouds, where the air temperature is below 0°C, are made up of tiny ice crystals, which join together to form snowflakes. The snowflakes fall and melt, becoming raindrops if the temperature of the air below is above 0°C; if the air below is around 0°C they fall as snowflakes. Snow on the ground is dry and powdery if the temperature is below 0°C but wetter if the temperature is around 0°C. In the picture the snow outdoors remains solid, but once it is brought into the café, on clothing, shoes and umbrellas, it melts. Other examples in the picture of materials in different states, which would change if they were not kept at the appropriate temperatures, are frozen foods, ice cream and butter. A useful change of state is the steam from the espresso coffee machine: steam has a higher temperature than hot water and brings out the flavour of the coffee more effectively.

Starting point

- Ask the children about places they know which have steamy or damp air. Introduce the words *humid* and *humidity*. Ask: *What causes this? Are these places always humid? What can be done to reduce the humidity? How does ventilation help?* Draw out that it lets damp air escape.

Using the picture

- The picture can help the children to assess their own learning about changes of state and the behaviour of materials in different states. Ask them to look for examples of water in different states and to explain why it is in solid, liquid or gas form in these parts of the picture.
- Ask: *Which solid materials in the picture would become liquid at room temperature and how are they being kept solid?*
- *Which gas is useful here and how is it produced?*
- Ask the children what they think the air in the café is like, and why. *Can they spot the causes of humidity?* (Melting and evaporating snow, steam from the coffee machine, people's breath.)

Activities

- The children could write a question about a material in the picture which changes state in cold or warm weather: for example, *What are the differences between ice, snow and hail? What makes hail fall when the temperature is above freezing point? What shape are snowflakes? Does snow melt more quickly than ice?*
- Help them to decide whether they can turn their question into a form which can be investigated practically or whether the answer can be found from secondary sources.
- Ask them to use all their investigations from topics 25 to 31 to help them to make a book about changes of state.

Higher-achieving children
Introduce the idea of materials being made up of tiny particles which we cannot see, which are continually moving about and are held together by invisible bonds. The children could enact the behaviour of particles, holding hands to represent the bonds. Ask: *Which has the most energy and has bonds which allow its particles to move about the most (gas)?* Ask them to enact its behaviour. They could do the same for liquid and solid and, at signals given by the teacher, enact different changes of state.

Lower-achieving children
Help them to collect a word bank of useful words for their 'Changes of state' book.

Plenary session

- Invite volunteers to read or show the class what they consider are the best parts of their 'Changes of state' book and to say what they found the most surprising.

Vocabulary

boil, Celsius (C), change of state, cold, condense, degrees, evaporate, freezing, frozen, gas, humid, humidity, ice, liquid, melting, snow, solid, steam, temperature, thermometer, water

Photo Book reference Page 31

28 The Earth from space

QCA links

The children learn that the Earth is approximately spherical, that it is sometimes difficult to collect evidence to test scientific ideas and that evidence may be indirect. (QCA Unit 5E)

Background information

It is often said that scientists in the past believed that the Earth is flat, but there is evidence from texts and maps, dating from about 150BCE or earlier, of the theory that the Earth is a sphere. Hellenic philosophers had started to look for rational rather than mythical explanations for their observations and explorers from the area had travelled as far as India, Afghanistan and the northern coasts of Norway.

Starting point

- Ask the children what they know about the Earth. Ask them to think about its shape, where it is, its movement and its relationship with other bodies in space. They could write their ideas on a concept map. Provide a base with some words to include (see Activities).

Using the picture

- Ask the children what they can find out about the Earth from the photograph: for example, shape, physical features, weather. Introduce the words *sphere* and *spherical*. *Can they name the large landmasses or identify any countries?*
- Discuss what surrounds the Earth and what is meant by *air* or *atmosphere*: *atmosphere* is the air (a mixture of gases) which surrounds the Earth. *What surrounds that?* Space contains no air; it is an empty space in which planets, stars and other bodies, such as meteorites, move.
- Take the children outdoors into an open place where they can see the horizon. Ask: *Does the world look like a sphere?* Ask them about the most distant objects they can see and introduce the term *horizon*. Discuss what happens to anything which moves into the distance: for example, a ship, a plane or a car. Why does it disappear? Using a tiny model ship with a football to represent the Earth, demonstrate that the ship moves out of sight as it follows the curve of the Earth's surface. Observations like this suggested to people long ago that the Earth was not flat, but they had no way of checking.

Activities

- The children could begin a word-processed book about the Earth, keying in information, scanning or downloading photographs, drawings and diagrams. They could build in hyperlinks to useful websites (see **Resources**).
- Ask them to include evidence which shows that the Earth is a sphere.
- ***Photocopiable sheet 11*** provides a base for a concept map about the Earth. Ask the children to write sentences using the words already provided and to add other words to link to them.

Higher-achieving children

Ask them to use their knowledge about the Earth and about forces to explain why people and objects do not fall off the Earth.

Lower-achieving children

Discuss photographs and paintings of the sea. Ask the children to point out the horizon. Ask: *Does the sea end there? Why does it stop there in the picture?*

Plenary session

- Draw out that we know what the Earth looks like from photographs taken from space, but that this was not possible until about 45 years ago; before that, people used other clues to find out about the shape of the Earth.

Vocabulary

air, atmosphere, continent, Earth, horizon, landmass, Pole, space, sphere, spherical

29 The Moon

Photo Book reference Page 32

QCA links
The children learn that the Moon is approximately spherical. They make observations of the Moon's apparent changes in shape which develop their understanding that the Moon takes approximately 28 days to orbit the Earth. (QCA Unit 5E)

Background information

The Moon is about a quarter of the size of the Earth, less dense (about 81 Moons would make up the same mass as the Earth) and has a much weaker gravitational force (about one sixth of that of the Earth). This is too weak to hold an atmosphere, so there is no air around the Moon and temperatures range from -150°C at night to 100°C at midday. It was thought that the Moon was completely arid, but evidence has been found of a small amount of water.

Starting point

■ Prepare for this topic a week or so beforehand. Prepare 30 yellow circles (12–16 cm diameter) to represent the shape of the Moon, and a strip of dark blue paper a little wider than the circles and long enough to make a 'Moon phase' chart. Mark the paper into segments large enough for a 'Moon'. For homework, ask the children to observe the Moon each evening and to draw its shape (they could first make 'Moon observation' note books). Begin a few days after the new moon and point out that it will not always be visible. It is useful for the teacher to make the same observations to check those made by the children. Each day, ask them to report briefly their observations. Introduce the terms *crescent* and *gibbous*. Discuss any discrepancies: for instance, which way the crescent faced. Each day, a different child could be responsible for cutting a yellow circle to represent the shape of the Moon, gluing it on to the appropriate place on the chart and writing the date.

Using the picture

❑ Ask the children to compare the photographs of the Earth (topic 28) and the Moon. *What similarities can they see?* (Both are spherical and both have light and dark areas.) *What are the dark areas on the surface of the Earth?* (Seas and oceans.) Explain that the dark areas on the Moon are flat expanses of dusty land; they were named *seas* because of the way they look from Earth.

❑ *What differences can the children see between the Earth and the Moon?* The Moon has no visible water; there are no clouds around the Moon. Clouds cannot form because there is virtually no water and no air.

Activities

- Ask the children to write notes about the Moon. *What else would they like to know?* Encourage them to write questions. Invite volunteers to read out a question. *Where can they find the answer?* The others could make suggestions. Ask: *Which books or CDs might help?* Bookmark any useful websites (see **Resources**).
- Ask the children what they observed about the cycle of changes. They could continue the observations and draw conclusions about them and about the time taken to complete the cycle. *Can they make links between this and the calendar?* Explain that long ago, people based the calendar on the phases of the Moon: some calendars have lunar months, but an actual lunar month has 28¼ days. A year is a little longer than 12 lunar months and the calendar has been adjusted to accommodate this.

Higher-achieving children
Ask them to find out about why the shape of the Moon appears to change and to use a model to explain this. They could use torches or lamps and balls of different sizes.

Lower-achieving children
Before discussing the Moon phases, set tasks in which they use a calendar to find out the number of days in each month, days in a year, and so on.

Plenary session

◆ Draw out that the calendar is based on observations of the Sun and Moon; the word *month* comes from *Moon* because the months are based on the phases of the Moon.

Vocabulary

atmosphere, calendar, crescent, Earth, gibbous, gravitational force, gravity, month, Moon, space, sphere, spherical, wane, wax

30 A Babylonian map of the world

Photo Book reference Page 16

QCA links
The children learn that sometimes it is difficult to collect evidence to test scientific ideas and that indirect evidence might have to be used. They learn that the interpretation of evidence depends on current scientific knowledge and understanding. (QCA Unit 5E)

Background information

The photograph shows a clay tablet from Babylon, dating from about 600 BCE. The drawing simplifies it and translates the text. People tend to draw maps with their own locality at the centre and work outwards from this point. Early map makers knew that something lay beyond the known areas of the world but they could only use theories about what was there. This map has Babylon (the capital of the ancient land of Mesopotamia – the part of Iraq between the rivers Tigris and Euphrates) at its centre because that was where it was drawn.

Starting points

- Ask the children to draw a sketch map of the locality of the school and to extend it outwards to include other places which they know are nearby. Invite them to compare their maps with those of others in the class and to notice any similarities and differences. *In what position on the paper did they place the school?*
- Provide large- and small-scale maps of the locality and the areas adjoining it and ask the children to compare them with their maps. *Which parts did they draw the most accurately?* Discuss the reasons for this. They are likely to know the areas immediately surrounding the school, and their routes to and from school, very well and know less about areas farther away.

Using the picture

- Do the children know the location of Babylon? Find out. Ask them to compare this with present-day maps of the world. Like modern maps, it shows that the Earth is round, but not necessarily that it is spherical. Point out that this might be because of a belief that the world was flat but that there is evidence that early civilisations had ideas that the Earth was a sphere. The children could look at modern maps and notice that on these the world is represented in a two-dimensional, rectangular form.
- *What does the map tell them about the Babylonians' knowledge of the world?* It looks as if they thought that the area they knew was the main part of the world and that there was one large landmass surrounded by an ocean, perhaps with other lands beyond it.

Activities

- Ask the children to imagine they were map makers in ancient times. *How would they find out what lay beyond the area they could see?* Draw out that they could collect reports from other people who have travelled in different directions. Point out that travel over land and sea was much slower in the distant past because the main sources of power were animals, humans and, for ships, the wind. This meant that the distances explorers travelled were limited, and sometimes they did not return.
- The children could compare this map with modern maps of the world and of Iraq. Ask: *Which cities might be represented by the circles?*

Higher-achieving children
The children could compare this map with later maps and comment on how knowledge of the world changed over time.

Lower-achieving children
Help them to locate the position of Babylon on a globe and to 'travel' with a finger from Babylon in different directions. *Which of these places can they find from the Babylonian map?*

Plenary session

- Draw out that map-making depends on observations and measurements. Point out that we now know about the main features of everywhere on Earth because new technologies, including satellite imaging, have helped but that there are still a few parts of the world which have not yet been mapped in detail.

Vocabulary

Babylon, Earth, sphere, spherical

Photo Book reference Page 34

31 A medieval map of the Solar System

QCA links

The children learn that it can be difficult to collect evidence to test scientific ides and that indirect evidence might have to be used. They also learn that the interpretation of evidence depends on current scientific knowledge and understanding. (QCA Unit 5E)

Background information

The photograph shows how medieval astronomers and scholars believed the solar system is arranged. Labels on the map include Terra (Earth), Ayre (air or atmosphere), Luna (Moon), Mercuris (Mercury), Sol (Sun), Saturnus (Saturn), Celstelat (sky and stars). Aristarchus of Samos, a Greek astronomer (c.310–230 BCE), was the first person known to have suggested that the Earth orbits the Sun. Because he had calculated that the Sun was much bigger than the Earth he thought it was unlikely that a large body would revolve around a smaller one. His work seems to have been ignored by later scientists, such as Ptolemy (85–165 CE), who put forward the theory of an Earth-centred universe.

Starting points

- For homework, ask the children to look at the sky on several evenings and to record the positions of the Moon, Venus (or other visible planets) and any stars they recognise. Discuss any changes. Discuss any previous work on the apparent movement of the Sun. Draw out that the Moon, Sun and any visible planets appear to move across the sky and that the stars gradually shift their positions over the course of the year; it looks as if they all revolve around the Earth but we know that only the Moon does so.
- Ask them what they know about the Solar System. Discuss what we can see: the Sun, Moon, Earth and stars. *Do they know the difference between a planet and a star?* Explain that a planet is a mass of material (some are made of rock but others are balls of gas with a core of rock); a star is a ball of burning gases with nuclear reactions taking place in its core, producing heat and light.

Using the picture

- ❑ Tell the children (or invite them to figure out) the meanings of the labels on the map. *What does it tell them about medieval knowledge of the Solar System – including its shape and boundaries?* It looks as if the map maker believed the Earth to be at the centre, with the Moon, planets and the Sun orbiting it. Discuss why (see Starting points). Draw out that, as we can see from observations, the Sun seems to move across the sky and looks as if it orbits the Earth. Tell the children about the theories of Aristarchus of Samos and Ptolemy and draw out that either the work of Aristarchus was lost or people did not believe his idea of a Sun-centred universe.
- ❑ Discuss the winged figures. Ask: *What are they and why should they be depicted?* Draw out that religion (Christianity in Britain) and science were very closely linked and that the angels suggest the idea of heaven surrounding the universe.
- ❑ Invite the children to name any planets they have heard of. List their responses. Ask: *Which planets are missing from the map, and why?* Draw out that the more distant planets could not be detected and so people did not know they existed.

Activities

- Ask the children to find out about the work of Copernicus and compare his Solar System map with this one.
- They could also find out about the discoveries of Uranus and Pluto, which are not shown on this map.

Higher-achieving children

They could find out about the theories of Copernicus and explain why the Christian Church was opposed to them.

Lower-achieving children

Help them to identify the main difference between Copernicus's universe and the one shown on this map.

Plenary session

- ◆ Draw out that it can be difficult to collect scientific evidence to explain an observation.

Vocabulary

Earth, Jupiter, Mercury, Moon, orbit, planet, Saturn, Solar System, star, Sun, Venus

32 A modern diagram of the Solar System

Photo Book reference Page 35

QCA links
The children learn that it can be difficult to collect evidence to test scientific ideas and that indirect evidence might have to be used. They also learn that the interpretation of evidence depends on current scientific knowledge and understanding. (QCA Unit 5E)

Background information

The Sun is the nearest star to the Earth. Nuclear reactions in the core of the Sun create heat and light, which support life on Earth. The planets orbit the Sun; their orbits are arranged in an approximately flat plane, except for that of Pluto, which is at an angle to the others. Some of them can be seen from Earth with the naked eye. Venus is the brightest and can be mistaken for a star; any 'star' which moves gradually across the sky is, in fact, a planet. Look west at least an hour after sunset or east just before sunrise.

Starting point

- Remind the children what they have learned about people's knowledge of space in the past and their theories for explaining their observations. Draw out that it was difficult to test the theories and that many of them could only be justified by argument. For example, Aristarchus of Samos argued that the Earth was more likely to orbit the Sun, rather than the other way round, because the Sun was much bigger than the Earth. Discuss the inventions which helped later astronomers to learn about space and discover new planets: telescopes, which became more and more powerful and, more recently, spacecraft which could collect observations using cameras and transport humans into space to make their own observations.

Using the picture

- Point out that the orbits of the planets have been drawn with 'cut outs' to show the planets in detail.
- Ask: *Which planets are the nearest to, and the farthest from, the Sun? What might it be like on these planets?* Make up mnemonics for remembering the order of the planets, beginning with the nearest to the Sun.
- Ask the children to compare this diagram with the medieval map of the Solar System. *What similarities and differences can they see? What has been learned?*
- Draw out the details of the modern map, such as the satellites (moons) orbiting other planets as well as the Earth. Ask: *Which other planets have moons?* The children could count the moons to find out which planet has the greatest number.

Activities

- Each group could research and make a factfile about a planet: distances from the Sun and Earth, mass, circumference, length of year and day, gravity in relation to that of the Earth, maximum and minimum temperatures, what it is made of and whether it has an atmosphere or water. They could write a weather forecast for their planet.
- Using information from the picture and their factfile, the children could make a model of the universe, from spheres such as beads, table tennis balls and other balls, to represent the movement of the planets, Sun and Moon (to an approximate scale).

Higher-achieving children
In their planetary weather forecast they could refer to the season and times of day (linking this to what they know about the length of the planet's year and day).

Lower-achieving children
Use models to help them to understand how a day and a year are linked to planetary rotation and orbit.

Plenary session

- Draw out that, like Earth, the other planets rotate and orbit the Sun.

Vocabulary

Earth, gravity, Jupiter, Mercury, Moon, Neptune, orbit, planet, Pluto, rotation, satellite, Saturn, Solar System, star, Sun, Venus

33 Sunrise and sunset

Photo Book reference Page 36

QCA links
The children learn that the Sun appears to move across the sky during the day (rising in the east and setting in the west), that the Earth's rotation takes 24 hours and that while some places have day others have night. (QCA Unit 5E)

Background information

The language used to describe the apparent movement of the Sun reflects the idea that the Sun, rather than the Earth, moves – the Sun *rises* (in the east) and *sets* (in the west). The Earth's rotation is the basis for the length of a day (including night). To measure time, the day has been divided into 24 segments (hours). In a complete rotation, the Earth spins through 360°, therefore in an hour it spins through $360 \div 24 = 15$ degrees (approximately). Each 15 degrees marks a time zone. Lines of longitude are marked in degrees, beginning with 0° at Greenwich, London – chosen (because of the observatory there) at an international meeting held in 1884 in Washington DC. (See ***Photocopiable sheet 12***.)

Starting point

■ Ask the children for words for the beginning and end of the day: for example, morning, sunrise, dawn, first light, evening, night, sunset, dusk, darkness.

Using the pictures

❑ Mask the captions and ask what time of the day each photograph shows. *How can we tell that neither of them shows midday?* (The Sun is too low in the sky). *Which is sunrise and which is sunset?* Explain that this is difficult without knowing the exact location of each place. *How would a map help?*

Activities

- The children could use a globe to find the longitude of London and other cities. Ask them to shine a torch on to the globe so that it illuminates the part of the world where they live and to turn the globe anti-clockwise. *Which parts of the world move into darkness (sunset) and for which does the 'Sun' rise?*

> ⚠ Warn the children not to look at the Sun; it can damage the eyes.

- Use the globe to show that the Earth's rotation determines the directions of sunrise and sunset and that time differs between east and west, not north and south. The children could find out where the Sun rises and sets in relation to landmarks. ***Photocopiable sheet 12*** explains how the Earth is split into segments, using lines of longitude, which converge at the Poles.

Higher-achieving children
They could predict (earlier or later), and then calculate from their longitude, the time of day in different places when it is, say, 7am, noon, 5pm and midnight in London.

Lower-achieving children
They could plan when to phone friends on holiday abroad, using a globe and torch to help them to check whether it will be day or night there.

Plenary session

◆ Draw out that, because the Earth rotates, daylight begins and ends at different times in different places.

Vocabulary

dawn, dusk, Earth, latitude, longitude, midday, midnight, noon, rise, rotation, segment, set, sunrise, sunset

34 The Earth's rotation

Photo Book reference Page 37

QCA links
The children learn that the Earth takes a year to make a complete orbit of the Sun. They present items of sunrise and sunset in a graph and learn to recognise trends and patterns in data. (QCA Unit 5E)

Background information

The Earth spins around an imaginary north-south line (axis) and is tilted at an angle of about 23° from the perpendicular to the plane on which it orbits the Sun; one hemisphere tilts towards the Sun and the other tilts away from it, creating warm and cold seasons. The temperatures of areas near the Equator differ little from season to season; this increases towards the Poles. In winter, the areas near a Pole have permanent darkness and, in summer, permanent daylight for about two months.

Starting point

- Ask the children during which season they stay outdoors the latest in the evenings, and why. Draw out that summer evenings are light and warm. *What do they know about daylight and weather in all four seasons?* Link this with work in geography. *What do they know about the seasons of another country, such as India?* Ask them to find India on a globe and to explain why its seasons differ from those of Britain.

Using the picture

- Remind the children about their previous learning about the Earth's rotation (see topic 32) and how it determines night and day. Ask: *In this illustration, for which part of the world is it day and for which part is it night?* Encourage use of the terms *east* and *west*. *For which part of the world will it soon be day? How can we tell?*
- Draw out that along the whole of an imaginary line around one side of the Earth, from one Pole to another, it is the same time of day. *Will daylight begin and end at the same time everywhere along this line?* Ask the children to point out the North and South Poles; point out that the Earth is slightly tilted (show them a globe and point out the tilt of the Earth). Ask them to use the picture, along with a globe and a battery lamp (which shines in all directions) to investigate the effects of the Earth's tilt as it spins and orbits the Sun, and to draw annotated diagrams to explain what happens.

Activities

- The children could use weather sections of local newspapers or the BBC weather website (www.bbc.co.uk/weather)to find out the times of sunrise and sunset for their locality. They could take turns to check this each day and to mark it on a line graph. *Do they think it will be the same for everywhere in their time zone?* (See topic 33). They could find out by emailing schools in the far north and far south of Britain and even in other countries. Other countries in the same time zone include France, Spain, Algeria, Mali, Burkina Faso and Ghana.
- Ask: *Is sunrise getting later or earlier? Is sunset getting later or earlier? What season are we in and which season are we moving towards?*

Higher-achieving children
Ask them to use the Internet to find sunset and sunrise times for towns on the same line of longitude and to record them on a line graph: for example, (around 19°E) Tromsø (Norway), Budapest (Hungary), Bangui (Central African Republic), Cape Town (South Africa). Different pairs of children could investigate other lines of longitude. *What pattern do they notice?*

Lower-achieving children
Fix a small piece of Blu-Tack to Britain on the globe and ask the children to notice if it has daylight or darkness as it rotates. Continue rotating the globe and move around the lantern to simulate the Earth's movement around the Sun. Stop at intervals and draw attention to the position of the Blu-Tack. *Is it on the part tilted towards or away from the Sun?*

Plenary session

- Draw out that places on the same line of longitude have the same time of day but that times of sunrise and sunset differ, depending on distance from the Poles.

Vocabulary

axis, Earth, east, Equator, hemisphere, longitude, North Pole, orbit, rotate, season, South Pole, west

Photo Book reference Page 38

35 An Inuit calendar

QCA links
The children learn that the Earth takes a year to make a complete orbit of the Sun. They learn to recognise trends and patterns in data. (QCA Unit 5E)

Background information

The seasons have different characteristics in different places, depending on physical features of the places, as well as latitude and longitude.

Starting point

- Ask the children what they know about seasons and what causes changes of season. *What differences have they discovered between the seasonal changes in different parts of the world?* Revise the terms *Equator*, *Poles* and *longitude*. Draw out that places near the Poles have greater variation in their seasons than places near the Equator.

Using the picture

- Explain that different peoples around the world developed different calendars to help them to measure time. The lengths of a year and a day are the same for everyone on Earth, but there have been different ways of splitting them into segments, such as months, days, hours, minutes and seconds, and different ways of distinguishing the seasons.
- Ask the children to discuss with a partner, and make notes about, what they can find out about the seasonal changes observed by the early Inuit who used this calendar. Ask: *Why did they have so many seasons? What differences and similarities are there between this and the calendar we use? How many months were in their calendar and how did they name them?*

Activities

- The children could invent calendars for their own locality and for a locality near the Equator, based on the seasonal changes about which they have learned. They could depict these on a circular format like that used to represent the Inuit calendar.
- Ask them to compare the calendars from different religions and find out about the localities where they originated. They could explain why the calendar from one locality might not be suitable for another.
- Draw out why the Universal Calendar (based on the Christian calendar) has been adopted as the official world calendar.

Higher-achieving children
Ask them to find out what a leap year is and why the Universal Calendar has them.

Lower-achieving children
They could cut out seasonal pictures from an old calendar and invent names for the months based on the natural events they observe. If possible, provide a calendar from another country, showing the weather and natural events. The children could compare the two.

Plenary session

- Draw out how the relationship between the Earth and the Sun affects people's ways of life and, in the past, the ways in which they developed calendars.

Vocabulary

axis, Earth, Equator, hemisphere, leap year, longitude, North Pole, orbit, rotate, season, South Pole, Sun, universal calendar

36 Sound spotter

Photo Book reference Page 39

QCA links
This can be used to introduce the topic of sound and to review what the children already know about how sounds are produced, about high and low, loud and quiet sounds. (QCA Unit 5F)

Background information

Whenever a sound is produced something moves and something vibrates; this could be a solid, liquid or a gas. We can hear several different sounds at once; some sounds can be distinguished more easily than others from the many sounds we hear; this could be because they are more familiar, more interesting, louder or higher than the others.

Starting point

- Ask the children to name some everyday objects which make sounds. Ask: *What kinds of sounds do they make?* Encourage the children to use a thesaurus to find the best words for the sounds. This could be linked with a literacy lesson.

Using the picture

- Working in groups, the children could list the objects in the picture which make sounds. Ask them to write the name of each object on a separate slip of paper and then to arrange the sounds in order, from quietest to loudest.
- They could copy their list and then arrange the sounds in order, from lowest to highest. Ask the children to make high and low sounds, then loud and quiet sounds, and discuss the difference between low and high and loud and quiet. Draw out that high sounds need not be loud. Ask the children to compare their two lists and invite feedback. *Were some of the high sounds quiet and some of the low sounds loud?*
- *Which sound do the children think they would hear the most clearly if they were in the street in the picture, and why?*

Activities

- *Which sounds do the children think they would hear in a local street? Which do they think they would find the most noticeable among all the other sounds? How could they find out?* Before going out to listen to the sounds in a local street tell the children to say, as soon as they get there, what sound they first notice. They could then list all the other sounds. Different groups could carry out this investigation at different times and then compare their results. *Were any sounds noticed particularly by all groups?* Discuss these results and help the children to explain them.
- They could plan a fair test to measure and compare the loudness of sounds in the street using a datalogger, which could present the results on graphs. Explain that the loudness of sound is measured in units called decibels (a decibel is a tenth of a bel – named after Alexander Graham Bell, the inventor of the telephone).

Higher-achieving children
Ask them to find out about the levels above which, and over what time period, sounds are considered to be a nuisance and what people can do about noise in the environment.

Lower-achieving children
Provide some practice in using a sound sensor with a datalogger. The children could practise using it to measure the intensity of sound from a radio as they alter the volume, watching the sound levels recorded on screen.

Plenary session

- Draw out that the sounds around us can have different levels of loudness and pitch and that we can measure and compare their volume.

Vocabulary

bang, bel, boom, clatter, click, clink, crackle, decibel, high, loud, low, piercing, pitch, quiet, rattle, ring, rumble, rustle, shrill, soft, sound, tap, vibrate, vibration, volume, whisper, whistle

Photo Book reference Page 40

QCA links
The children learn that sounds are made when objects or materials vibrate; 'pitch' describes how high or low a sound is and that the pitch of a stringed instrument depends on the length and thickness of the strings. They have opportunities to suggest how to change the pitch and loudness of stringed instruments. (QCA Unit 5F)

37 A harp

Background information

High sounds are produced when something vibrates at a high frequency. Low sounds are produced when it vibrates at a low frequency. In general, small objects or small areas of a material vibrate at a higher frequency than large objects or large areas of a material.

Starting points

- Set up a series of pitch investigations for the children to try. Ask them to make a note of how the sound was produced and to draw the objects in order: lowest to highest pitch.
  1. Suspend from strings three ceramic plant pots, of different sizes, or iron bars or pieces of metal pipe of different lengths, and invite the children to tap these objects with a metal tapper.
  2. Knock nails of different lengths and thicknesses into a solid block of wood and ask the children to tap them with a very large nail.
  3. Pour water to different levels in 4 identical bottles and ask the children:
     a. to tap them, and
     b. to blow across the tops of them.
  4. Provide drums of different sizes for the children to tap with a drumstick.
- *What did the children notice about the objects which made the lowest pitched sounds? What did they notice about the ones which made the highest pitched sounds?* Discuss what vibrated in each investigation (when the bottles are tapped the bottle and water vibrate; when the children blow across the tops of them the column of air inside them vibrates).

Using the picture

- Ask the children to make notes about the type of sound made by the harp and how the sound is produced. Ask: *What vibrates?*
- *Which string do they think produces the lowest pitched sound and which produces the highest sound?* Ask them to explain their answers in relation to what they have learned from their investigations of pitch.

Activities

- The children could plan an investigation to test their ideas about the sounds produced by strings (or elastic bands) of different lengths.
- They could stretch elastic bands of different lengths across different-sized boxes or tins and pluck the elastic bands. Ask them to draw and label the boxes or tins in order, from lowest to highest sound. *What pattern do the children notice?*

Higher-achieving children
Ask them to find out how the pitch of sounds is varied in instruments whose strings are the same lengths: for example, guitar, violin.

Lower-achieving children
Provide practice in recognising high and low sounds before the children begin the investigation. They could represent them on a ladder – objects which produce the lowest sounds at the bottom and progressing up the ladder to those which produce the highest sounds at the top.

Plenary session

- Draw out that smaller objects and areas of material (including air) vibrate more quickly than larger ones and produce higher sounds.

Vocabulary

high, low, pitch, vibrate

38 A stethoscope

Photo Book reference Page 41

QCA links
The children learn that sounds can travel from a source, through solids, liquids and gases, by making the materials vibrate. (QCA Unit 5F)

Background information

On one side of the end of the stethoscope is a flat, thin membrane. Sounds from the chest make it vibrate and this makes the tubes vibrate. The other side (the *bell*) has no membrane. It is placed lightly on the skin and uses the skin as a membrane. The bell picks up lower-pitched sounds and the membrane picks up higher-pitched sounds.

Starting points

- Ask the children if sounds travel through air. *How do they know?* Use everyday examples – we can hear people speaking, a bell ringing, a dog barking, and so on. *Do they think sounds travel through water? Can they hear anything when their head is under water?* Find out during swimming lessons. *Do they think sounds travel through solid materials, such as wood or metal?* They can find out by tapping a tabletop and listening, both with their ear on and off the table. Ask: *Which sounds louder?* Tap on a water pipe and ask them to listen, first with their ear off the pipe and then on it; repeat this, increasing the distance between the source of the sound and the listener (if possible find a pipe which extends out of the classroom).
- Discuss how the sounds travel through the air, water and wood or metal (the tapping, or other sound, is produced by a vibrating material or object, which makes materials or objects touching it vibrate).

Using the picture

- Ask the children to name the instrument in the photograph. Ask: *What might the doctor hear when she listens to the girl's chest using the stethoscope? Can we hear someone's heartbeat without a stethoscope? What difference does a stethoscope make?*
- Ask the children to look at the stethoscope carefully and, remembering what they have discovered about how sounds travel, to explain how they think it works. Explain that the head of the stethoscope can be turned over. Introduce the term *membrane* and explain that the side touching the girl's chest is flat, with a membrane which vibrates easily, especially for high-pitched sounds. Ask: *What makes it vibrate and what happens when the membrane vibrates?* Point out the cone-shaped side (the *bell*) and explain that this side can be placed on the patient's chest to listen to lower-pitched sounds. *What do the children think happens to help the doctor to hear the sounds?*

Activities

- The children could think of a way of making a stethoscope to listen to one another's heartbeat. They could make a single earpiece and listening piece from funnels or pierced yogurt pots, linked by a piece of rubber or plastic tubing. A stethoscope with a double earpiece can be made using two pieces of plastic tubing linked to a third piece by a T-shaped junction (available from laboratory suppliers).
- They could also make a 'telephone' from cans or yogurt pots and string and compare the ways in which this and the stethoscope work.

Higher-achieving children
Challenge them to plan an investigation to find out if a tube can help them to hear sounds from longer distances (they could use a long hosepipe with funnels attached).

Lower-achieving children
Show them examples of membranes which vibrate: for example, a drum, a tambourine. Place rice on the drum or tambourine and let them watch what happens when the instrument is struck lightly.

Plenary session

- Draw out that sounds travel more easily through solids than through air, and that they travel more easily through air within tubes than through larger areas of air.

Vocabulary

hollow, membrane, stethoscope, vibrate

Photo Book reference Page 42

QCA links
The children learn that vibrations from sound sources travel through different materials to the ear and that we hear them because they make objects and materials in our ears vibrate. (QCA Unit 5F)

39 How we hear sounds

Background information

Sound sources vibrate, making the air vibrate; vibrating air makes the eardrum (a membrane in the ear canal) vibrate and this, in turn, causes vibrations in the ossicles (small bones known as the hammer, anvil and stirrup). These bones vibrate against the cochlea (a spiral fluid-filled tube in the form of a cone – rather like a snail's shell); pressure changes in the fluid trigger nerve impulses, which travel along the auditory nerve to the brain, which interprets them. The semicircular canals are the organs of balance.

Starting point

- Ask the children how they think we hear sounds. They could draw a diagram to explain what they think happens. Ask them how the shape of the ear helps to collect sounds (compare it with the shape of a funnel attached to a tube).

Using the picture

- Ask the children to compare this diagram with the ones they drew to explain how we hear sounds. *Which parts could they not explain very well, and why?* Point out that this diagram shows the internal part of the ear, which we cannot see.
- Help them to trace the course of the vibrations caused by vibrating air which is caught by the ear lobe and channelled down the ear canal. *Which solid object does this make vibrate?* Point out the other solid objects which, in turn, begin to vibrate. Draw attention to the cochlea, in which a fluid vibrates.

Activities

- The children could explore ways of transmitting sound vibrations through different materials. Challenge them to transmit a sound (such as the tapping of two spoons or two nails against one another) across the room, through at least three different materials. *Can the sound be heard at the other side of the room? What is the greatest distance across which the children can transmit this sound? Which materials were the most effective?* They could use tubes, hosepipes, funnels, cans, metal bars and pipes, a tank of water or a water-filled pipe.
- They could try transmitting a sound across the room, through four or five different materials.
- Ask them to use Morse code (see **Resources**) to send messages to one another via the 'sound routes' they have created.

Higher-achieving children
Challenge them to make a model of the human ear, which receives sounds and transmits them from one material to another, in a similar way to a real ear.

Lower-achieving children
Remind them of their investigations of sounds passing through solid materials (see topic 37) and discuss how this can help them to transmit sounds across the room.

Plenary session

- Draw out that sounds can travel long distances through some materials. Ask the children what they notice about the materials which conduct sounds well.

Vocabulary

brain, cochlea, ear canal, eardrum, ear lobe, membrane, nerve, stethoscope, vibrate

Photo Book reference Page 43

40 Helen Keller

QCA links
The children learn that we can see and feel the vibrations of sounds. (QCA Unit 5F)

Background information

Anything which produces or transmits a sound vibrates. Thomas Edison is credited with inventing the phonograph (1877). He had an idea that sounds had individual 'shapes' which could be recorded. He used a vibrating stylus to make indentations in tinfoil wrapped around a rotating cylinder, which could be played on a phonograph. Helen Keller was blind and deaf from the age of 19 months. She could not speak until Anne Sullivan taught her to communicate through feeling the vibrations of people's throats and the movements of their lips, and to make the same sounds herself.

Starting point

- Ask another adult to put on a blindfold and a pair of soundproof earmuffs. Ask the children if this person will be able to tell if the radio is switched on. *How? What could he or she do to find out (apart from removing the earmuffs or opening his or her eyes)?*

Using the picture

- Tell the children about Helen Keller (see Background information). Ask: *How can she hear the phonograph?* Remind them about what they know about how sounds are made and transmitted from a source to the ears.
- *Can they now think of a way for a person wearing a blindfold and earmuffs to tell when a radio is switched on?* Ask them to write instructions for him or her to try.

Activities

- Provide a collection of musical instruments and ask the children to work in groups; one child wears earmuffs and a blindfold while the others make a sound with an instrument. He or she should touch the instrument and describe what can be felt. *Can he or she tell when the instrument is making a sound? Is it possible to distinguish between loud and quiet or high and low sounds through the sense of touch? Does this become easier with practice?*
- ***Photocopiable sheet 13*** provides a 'circus' of activities which show the vibrations caused by sounds. Ask the children to describe and explain their observations of each activity.

Higher-achieving children
Ask them to suggest a way of making vibrations visible, so that someone who cannot hear knows when a sound is made.

Lower-achieving children
Provide a word bank to help them to label their diagrams

Plenary session

- Draw out that anything which makes or transmits a sound vibrates and that we can sometimes see these vibrations.

Vocabulary

phonograph, stylus, transmit, vibrate

41 Alexander Graham Bell's first telephone

Photo Book reference Page 44

QCA links
The children learn that sounds are made when objects vibrate and that sounds from various sources travel, through different materials, to the ear. (QCA Unit 5F)

Background information

Alexander Graham Bell invented the telephone in 1875, with the help of Thomas Watson. To use the telephone, Bell spoke into a hole in the bottom of the telephone and the sound made a flexible diaphragm vibrate. The diaphragm was attached to the end of a steel reed positioned over the pole of an electromagnet. The other end of the reed was hinged to the opposite pole of the same magnet. This caused tiny variations in the electric current travelling down a wire to a reed receiver. Recognisable sounds could be heard but the voice was not heard distinctly. The telephone was gradually improved until the voice could be heard clearly.

Starting point

- Remind the children about the ways in which they tapped out Morse code messages (see topic 39). Tell them that at one time this was the only way in which messages could be transmitted along wires. There was a limit to the distance the messages could travel and operators, who received them on machines which tapped out the Morse code letters, had to re-key them along the route. The final message had to be transcribed by an operator. It was much quicker than any known method, but was a very expensive way of communicating and so was used only for very important or urgent messages.

Using the pictures

- Cover the captions and other text on the photograph and diagram and ask the children if they have any idea what they show. Let them uncover the captions and text and ask them if they have any idea how this telephone could transmit sounds. *What do they know about how sounds are made and transmitted?* Tell them that the user had to speak into the bottom of this device. *What do they think might vibrate? Where might the vibrations be transmitted? Can they see anything which looks like a membrane?* Introduce the word diaphragm for a membrane or thin skin.
- They could work in pairs to write an explanation of the ways in which vibrations travel from this telephone to another one.

Activities

- The children could investigate the ways in which sounds, such as the human voice, can make a diaphragm vibrate:

Making a diaphragm, or membrane, vibrate

- Ask them to observe what happens when they speak more quietly or more loudly and to write a report about their investigation.

Higher-achieving children
They could also make an electromagnet and find a way of using it to tap out Morse code messages (see **Resources**).

Lower-achieving children
Make an enlarged copy of the diagram (not the photograph) and ask the children to draw arrows to show the direction in which the sound of the voice travels from the base of Bell's telephone.

Plenary session

- Draw out that the first telephone made use of vibrating air caused by a human voice.

Vocabulary

diaphragm, electromagnet, membrane, Morse code, vibration

42 Recording sound in the past

Photo Book reference Page 45

QCA links
The children learn that sounds are made when objects vibrate, that sounds from various sources travel through different materials to the ear and that sounds can be recorded. (QCA Unit 5F)

Background information

Thomas Edison invented the phonograph in 1877 (see topic 40). The sounds were collected by a long, hollow horn and their vibrations were transferred to a needle, which cut tracks on a cylinder. The louder the sound, the deeper the indentation. The top picture shows a cylinder being cut for a phonograph. Emil Berliner found a method, in 1887, of using a needle in a similar way as Edison, but to trace a spiral sound groove on a flat disc instead of a cylinder. A negative was made from the disc and used as a mould for making multiple copies. These *records*, as they came to be known, could be played on a device which Berliner named a *gramophone*. The original gramophones were similar to the one in the picture. A stylus followed the groove of the record, picking up the sounds; their vibrations were amplified by the horn with its hollow tube.

Starting point

- Return to the photograph of Helen Keller listening to a phonograph (topic 40). *Do the children remember how the sounds were recorded and played?* Tell them that another inventor, Emil Berliner, found a different way of recording sounds so that copies could be made of the recording.

Using the pictures

- Ask the children to work in pairs to describe how the cylinder for a phonograph is being recorded in the top picture. Point out that the sounds from the piano make the air around them vibrate and that some of these vibrations reach the trumpet-like device held close to the piano. Ask: *Why has this device such a large, trumpet-shaped end?*
- They could also work in pairs to write an explanation of how the gramophone in the bottom-right picture works. *Where does the sound come out?* Ask them to explain why the gramophone has a large horn.

Activities

- Ask the children to plan an investigation to find out how a trumpet shape could help them to hear. They could make different-sized 'ear trumpets' from card and compare the effects of using them as a hearing aid:

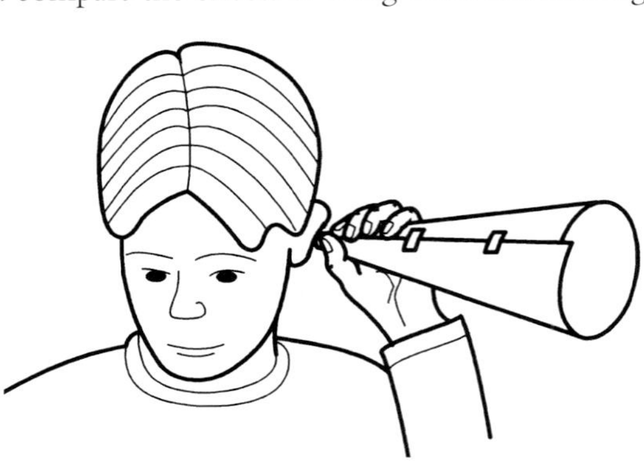

- Ask them how they can measure the loudness of the sounds collected by the 'hearing aids'. Remind them of the ways in which a datalogger and sound sensor can be used to measure the loudness of sounds and ask them to present the results in the form of a graph.

Higher-achieving children
They could also investigate a home-made megaphone to magnify the sound of their voice.

Lower-achieving children
They could first practise using a sound sensor to record sounds they make: ask them to make loud and quiet sounds and notice the difference on the graph.

Plenary session

- Draw out that sounds were first recorded by turning vibrations into shapes and that these shapes could be converted back into the original sounds.

Vocabulary

cylinder, disc, gramophone, groove, megaphone, phonograph, record, spiral, stylus, vibration

43 Walls

Photo Book reference Page 46

QCA links
The children learn to ask scientific questions and to plan how to answer them. (QCA Unit 5H)

Background information

The photographs were taken around the same school on the same day. The children used a compass to check the directions in which the walls faced. They were encouraged to notice the differences between the way in which the weather had affected the walls and the numbers and types of plants growing on or in front of them.

Starting point

- Ask the children what they would expect to see if they examined the exterior walls of the school or playground. List their responses. Ask: *Are there any living things on the list?*

Using the pictures

- Ask the children to work in pairs to describe the different walls around the school in the pictures. *What similarities and differences do they notice?* All the walls are made of a similar material (brick), but some of them have plants growing on them or on the ground in front of them, and there are different types of plants.
- Ask them to write explanations about why some walls have plants growing on or near them but others do not, and why the north-facing wall might have a small patch of moss only in one place and nowhere else.

Activities

- Take the children to look at the walls around the school. Provide compasses so that they can check the directions in which the walls face (they could compare different sides of the same wall).
- Provide a digital camera so that they can photograph the walls and take close-ups of any plants for later identification using information books and pictures from various sources, including the Internet and local field guides.
- Encourage them to ask questions which could later be turned into investigations to help them to explain any differences between plant-growth in different places on walls: for example, *Has this wall been cleaned or painted? Do we play near this wall? Which way does the wind usually blow? In which direction does the Sun shine most of the time? Is the wall in a sheltered place? Is it usually wet or dry? Which direction does most rain come from? What conditions help most plants to grow? What conditions does this plant prefer?*

Higher-achieving children
Encourage them to find out in which places around the school moss grows, and why. They could find out if this is true for other schools and set up email links with them which could also be used by others in the class to check their predictions about plant growth on walls of other schools and in different parts of the country.

Lower-achieving children
Help them to organise their observations on a chart: for example, with a section for each wall, headed 'north-facing', 'south-facing', and so on. They might need practice in using a compass.

Plenary session

- Draw out that walls which face different directions can have different weather conditions and can become habitats for a range of living things.

Vocabulary

compass, direction, east, fern, habitat, moss, north, south, west

44 Plants on waste ground

Background information

The plants in the photograph grew on a patch of poor soil where part of a building was demolished. They were not planted there. Some grew from seeds or roots which were already in the soil; other seeds were blown there on the wind or carried by animals.

Starting point

■ Discuss why some plants are called *weeds*. Ask: *How do we know if a plant is a weed?* Draw out that this term refers to any plants which grow where they are not wanted: for example, some people plant poppies in their gardens but farmers who grow crops might regard them as weeds; there are gardeners who plant foxgloves, which also grow wild in hedgerows; some people value dandelions because their leaves can be eaten in salads but many gardeners remove them from their lawns because they consider them to be weeds.

Using the picture

- ❑ Ask the children if the plants in the picture are weeds and to explain their answers.
- ❑ Tell them that no one planted any of them; they just appeared on this piece of ground where there used to be a building. *How could they have got there?*
- ❑ Ask the children to list their ideas and then invite feedback. Help them to turn their ideas into forms which can be investigated: for example, *What will happen if we clear a square of soil and then leave it without doing anything to it? Will the same happen if we bring some soil indoors and spread it over an area of the same size? How can we find if plants grow because the seeds or roots are already in the ground or if they come from somewhere else?*

Photo Book reference Page 47

QCA links
The children have opportunities to decide what evidence to collect in order to answer scientific questions and to collect and record data appropriately. (QCA Unit 5H)

Activities

- Different groups could investigate a cleared area of soil which is treated in different ways. To keep out airborne seeds one patch could be covered with a clear plastic tonneau (available from garden centres). Spread some soil in a seed tray or the base of a propagator and keep it indoors to stop any seeds landing on it. To make the test fair, the soil kept indoors should be watered. A control patch of soil of the same area should be cleared outdoors and given no treatment. Ask the children to predict what will happen to each patch of soil and what this result will tell them. *On which patches of soil do they think the most plants will grow, and why?*
- They could photograph the soil each week, make a note of the height of any plants which grow and identify them, where possible.

Higher-achieving children
They could plan an investigation to find out which plants tend to choke out others (and for this reason tend to be treated as weeds).

Lower-achieving children
Help them to recognise common plants such as bindweed, clover, daisy, dandelion, dock, foxglove, nettle, plantain.

Plenary session

◆ Draw out that plants appear in soils because their seeds are carried there, their roots spread under ground or over the surface, and so on – they do not appear 'from nowhere'. The children should have discovered that most soil contains plant material. Draw out that some plants can tolerate a greater range of conditions than others and that this is why they are very common.

Vocabulary

disperse, root, seed, shoot, soil, weed

45 Plants from nowhere

Photo Book reference Page 48

QCA links
The children have opportunities to collect and record data appropriately, identify and describe patterns in data, look critically at data collected and try to explain their results using their scientific knowledge and understanding. (QCA Unit 5H)

Background information

The plants in both photographs are rosebay willowherb, which grows well on cultivated land and wasteland, woods, ditches and hedgerow; especially where the ground has been disturbed. It is commonly known as *fireweed* because of its tendency to colonise areas where there has been a fire. The heat and smoke of a fire has been found to encourage the germination of its seeds, which can lie dormant for several years. It has numerous tiny brown seeds (about 380 per pod and 80,000 per plant), which have silky hairs to aid wind dispersal. Rosebay willowherb, which can grow to a height of 150cm, often becomes a weed and a dominant species on disturbed ground. Once established, the plants spread extensively underground by means of their roots. It is a very pervasive plant which forms very dense clumps, excluding other plants.

Starting points

■ The children should first have completed topic 44. *What did they find out about the plants which appear in cleared patches of soil? Which were the most common ones?* Discuss why these plants appear the most often in cleared patches of soil: for example, they might already be established nearby, produce numerous or easily dispersed seeds, be particularly well-suited to the habitat, spread rapidly by means of roots and tolerate a variety of conditions.

Using the pictures

❑ Discuss the appearance of the plant and ask the children to describe it in a way which helps others to recognise it: for example, approximate height, flowers, leaves, stem, ground cover.

❑ What do they notice about the places where the plant is growing? Are they wild places or have they been cultivated? Has anything else happened here?

❑ *Have the children seen rosebay willowherb before? Where?* Discuss why it is known as *fireweed. Have the children seen it growing in places where there have been fires?*

> ⚠ An adult should light the fire. Show the children how to prevent it spreading by surrounding it with stones or bricks.

Activities

- Take the children out on a 'rosebay willowherb safari', making notes about the places and conditions where they see the plant growing: for example, cultivated land, waste ground, roadside verges, woodland borders, hedgerows, places where there is rubbish or there has been a fire. Use a digital camera to photograph these places. Draw out that rosebay willowherb grows in a variety of conditions.
- Discuss how to find out what encourages rosebay willowherb to grow on a patch of ground. Find an area which is not far from where it already grows, or transplant some to a site near where the children can carry out their investigation.
- Different groups could treat areas of ground (of the same size) in different ways. Mark out areas of $1m^2$: one with grass and other small plants growing on it (do not treat it in any way); clear another area but do not treat it in any other way, clear and dig over another area, on another area light a small fire of wood shavings and small pieces of wood (keep it burning for as long as possible). The children could mark the areas into squares using string pegged to the ground, make a copy (to scale) on squared paper and record whether fireweed grows in each square.

Higher-achieving children
They could plan an investigation to find out if heat (without fire) encourages rosebay willowherb seeds to germinate more quickly.

Lower-achieving children
Show them how to use squared paper for making a scaled plan of the area of ground they are observing.

Plenary session

◆ Invite the children to report their results and discuss how these results can be interpreted.

Vocabulary

cultivated, disperse, fireweed, root, rosebay willowherb, seed, shoot, soil, waste ground, weed

Photocopiable sheet 1

Heart and pulse

You can feel a pulse beat each time your heart beats.

You need a stopwatch.

- Feel your pulse in one of these places.

the crook of an elbow

behind a knee

on your neck

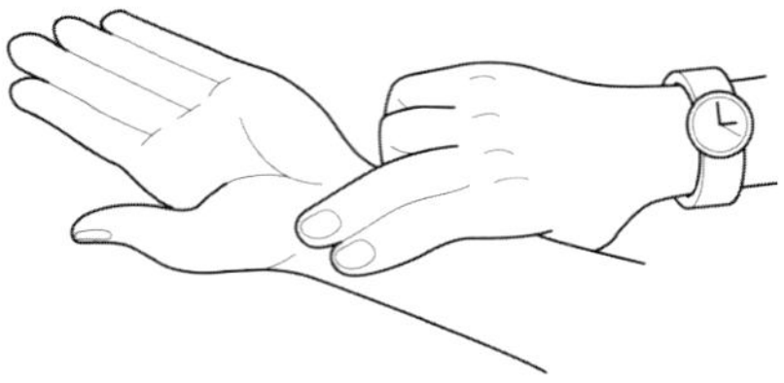

your wrist

- Count your pulse beat while a friend times 30 seconds.
- Record your pulse rate after walking and running for different lengths of time.

Do not press on pulse spots.

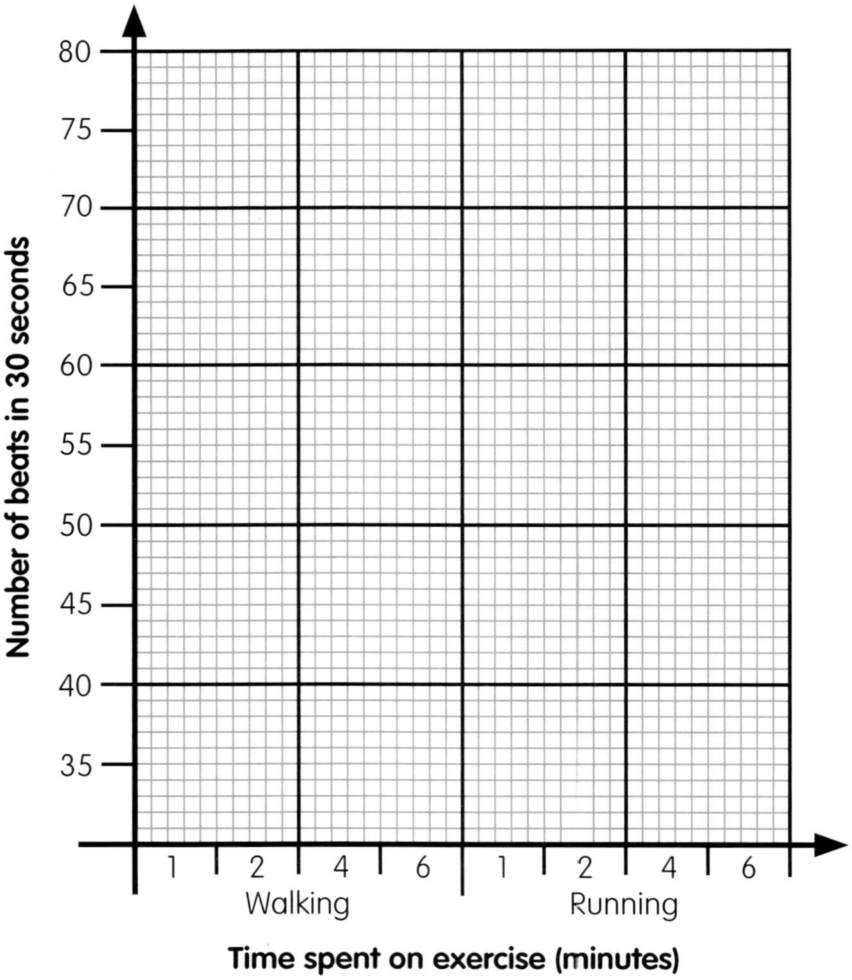

Which exercise raised your pulse rate the most?

____

____

____

____

____

____

____

____

Pollen power

- Fill in the labels to explain how a plant is pollinated.

useful words

female	ovule
fertilise	pollen
male	stamen
ovary	stigma

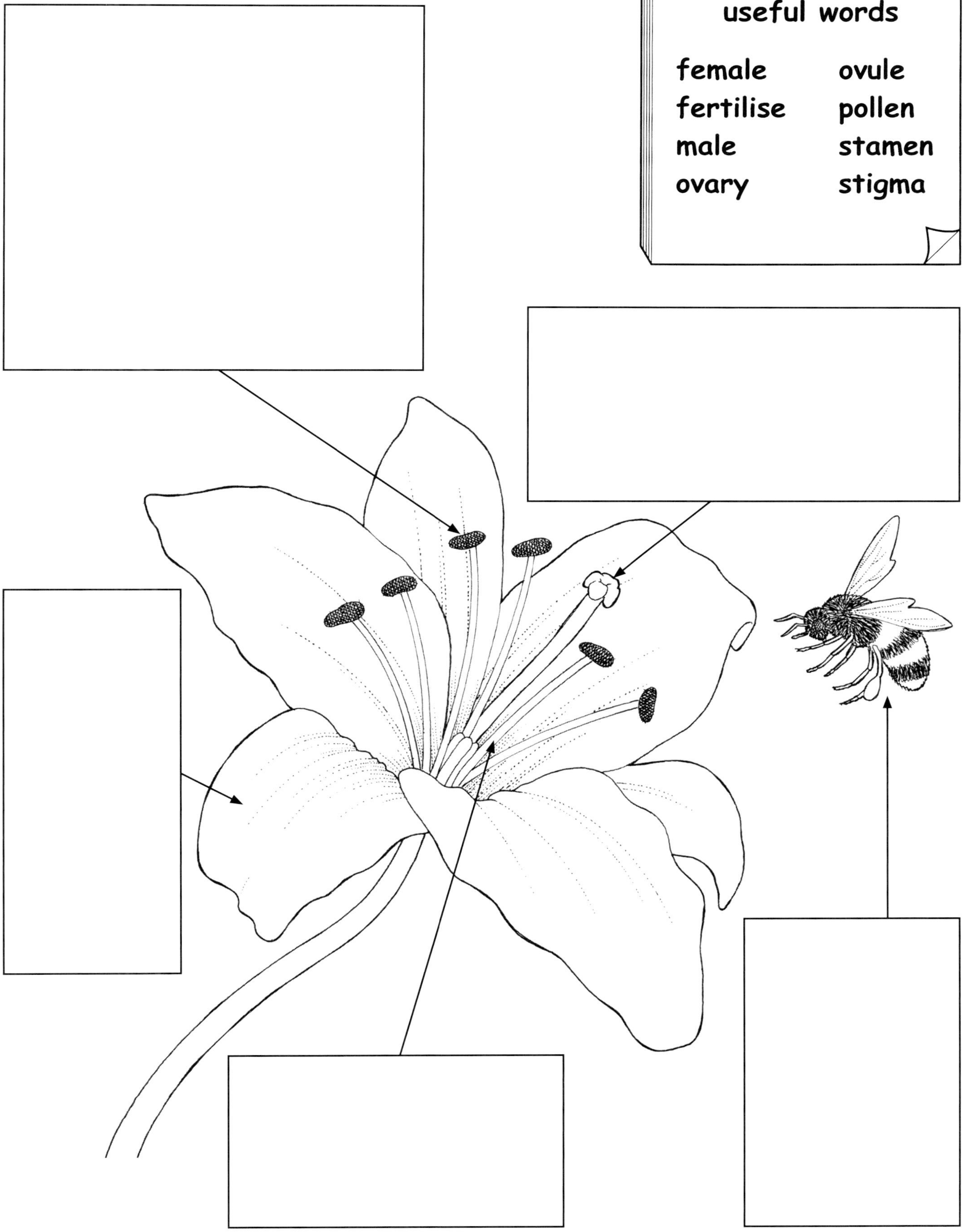

Photocopiable sheet 3

Soaking sponge

Which soaks up more water:

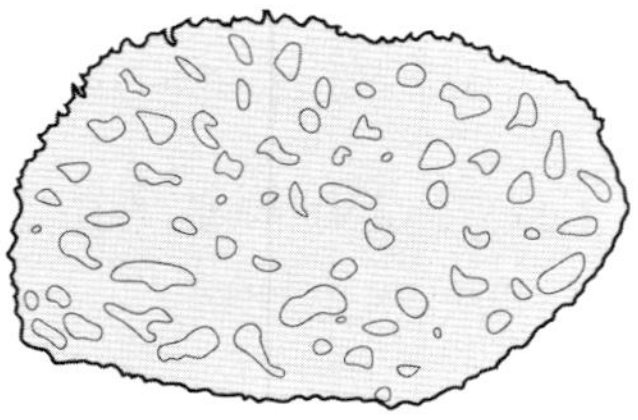

natural sponge

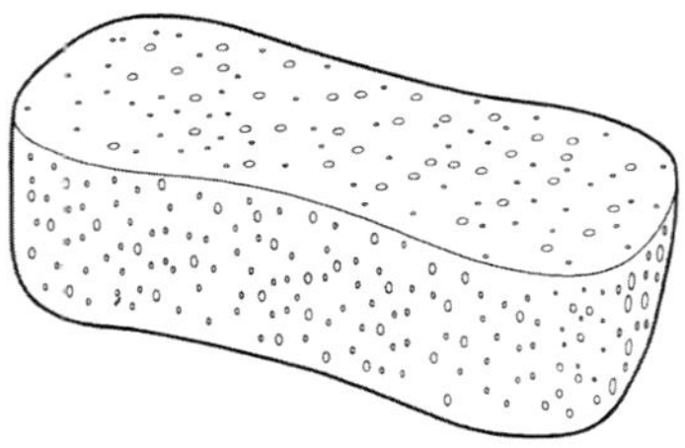

synthetic sponge

or is there no difference?

You need

natural and synthetic sponges of the same size, water, measuring jug.

1. Put 1 litre of water in a measuring jug.
2. Push a sponge into the water.
3. Take out the sponge. Do not squeeze it.

- Record your results on the chart.

Type of sponge	Volume of water		
	Before	**After**	**Soaked up by sponge**
Natural			
Synthetic			
Natural			
Synthetic			
Natural			
Synthetic			

Each pair of sponges should be the same size.

Conclusion ______________________________

____

As light as air

Does air in an object make it lighter or heavier?

Prediction ____________________

You need

4 balloons of the same size, balloon pump, a thin piece of dowel, string, ruler

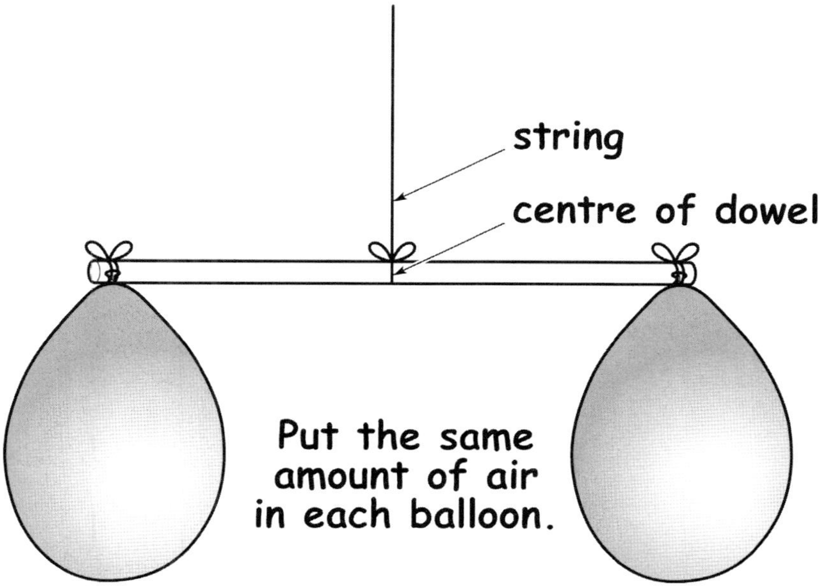

⚠ Always use a balloon pump.

- Repeat this, but inflate only one balloon. Draw what happens.

- Explain your result ____

____

Was your prediction correctt? ____________________

Does air make a balloon lighter or heavier? ____________________

Strong pong

- Find out if some perfumes have stronger smells than others.

Work with a group.
One person sprays 3 squirts of perfume on to a piece of kitchen paper.

The others should be at least 4 metres away.

You need

3 different spray perfumes or aftershave lotions, a long tape measure and kitchen paper.

- Walk towards the perfume. Stop when you can smell it.

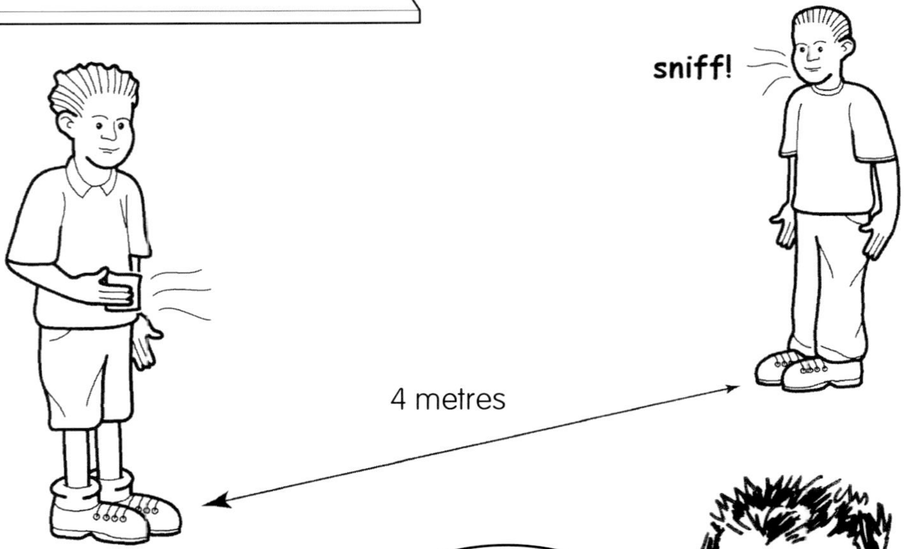

- Measure the distance between you and the perfume.
- Record the results for each tester.

Name of perfume	Smelling distance			
	Tester 1	Tester 2	Tester 3	Tester 4

Conclusion ____

____

Spray works

- Fill in the labels to explain how a spray bottle works.

Useful words

force	reservoir
lever	spring
liquid	squeeze
nozzle	suck
pressure	tube
pump	valve

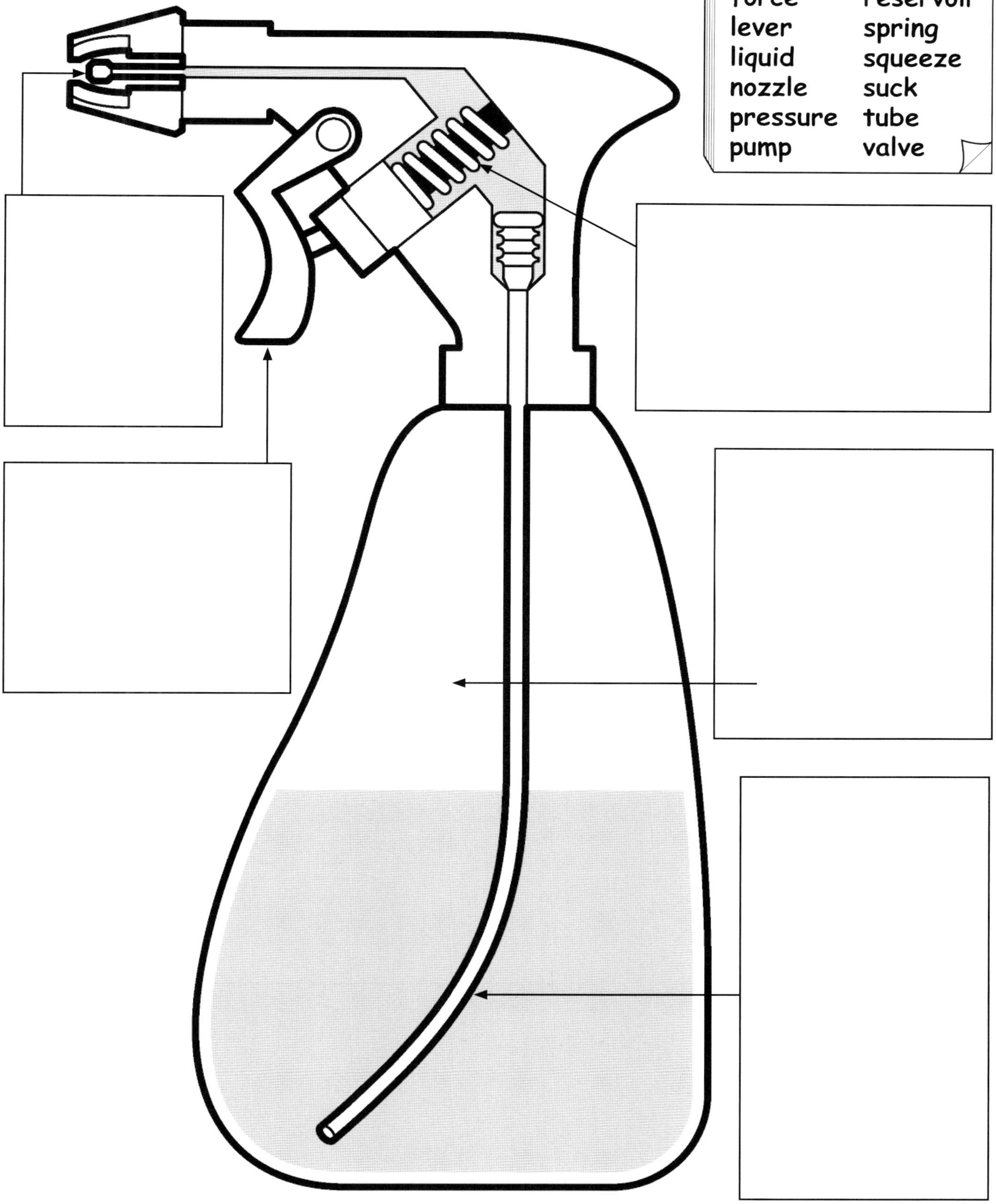

Waterlogged

You need

3 different types of soil: clay, sand, loam
a magnifying glass,
a teaspoon,
a measuring cylinder.

- Use a magnifying glass to observe the soils.

Which do you think contains the most air? ______________________

Why? ______________________

____

____

Water soaks into soil which has air holes.

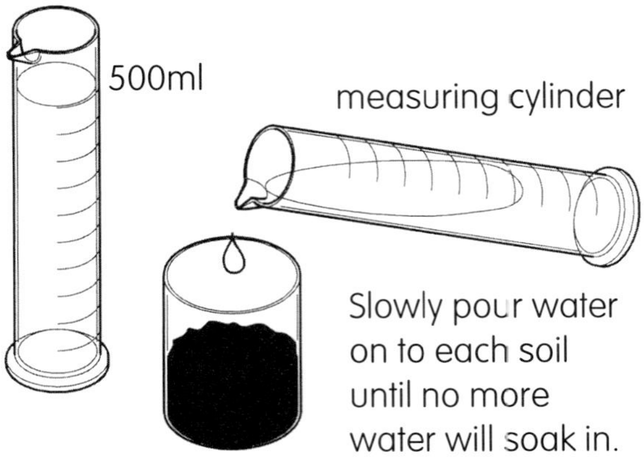

Slowly pour water on to each soil until no more water will soak in.

- Record your results.

Soil type	Volume of water in measuring cylinder		Volume of water which soaked into soil	Observations
	Before	After		
clay				
sand				
loam				

Conclusion ______________________

____

Rain from the sea

Water from the sea evaporates into the air. When the air cools, the water vapour condenses. Does it produce salt water rain?

- Plan an investigation that you can carry out at school.

Is rain from sea water salty?

What we shall do

What we need

Health and safety

Our prediction

What will happen if our prediction is correct

Results

Conclusion

Why this happened

Solid, liquid and gas game

For each line of this game, choose a letter at random.

- Think of a solid, a liquid and a gas beginning with that letter.

  The first line has been filled in for you.

Letter	Solid	Liquid	Gas	Score *
N	nickel	nail varnish	nitrogen	

*** Scoring**

If no one else has the same material	6	If 3 others have it	3
If 1 other person has it	5	If 4 others have it	2
If 2 others have it	4	If 5 or more others have it	1

Drying

Which is the quickest way to dry the washing?

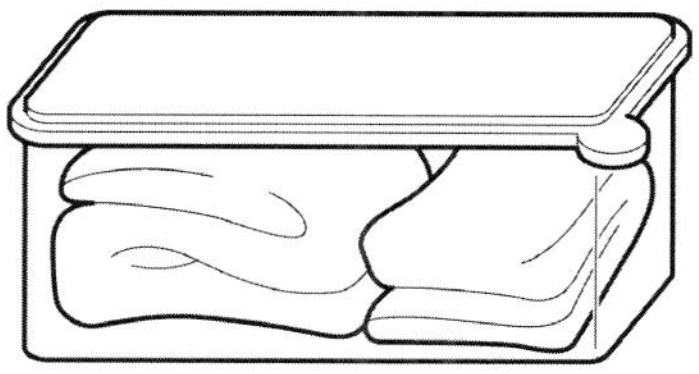

in an airtight box ☐

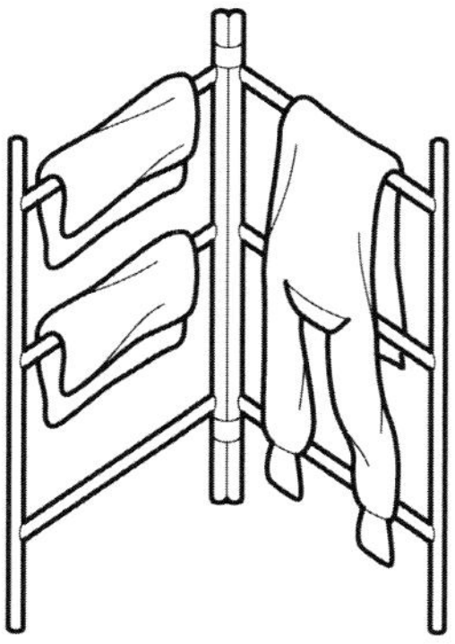

on an indoor airer ☐

on an outdoor clothes line ☐

in a tumble dryer ☐

- Explain your answer ______________________________

- Plan a fair test to find out.

What we shall do	What we need
	Health and safety 
How we shall make the test fair	

Photocopiable sheet 11

Earth concept map

What do you know about the Earth?

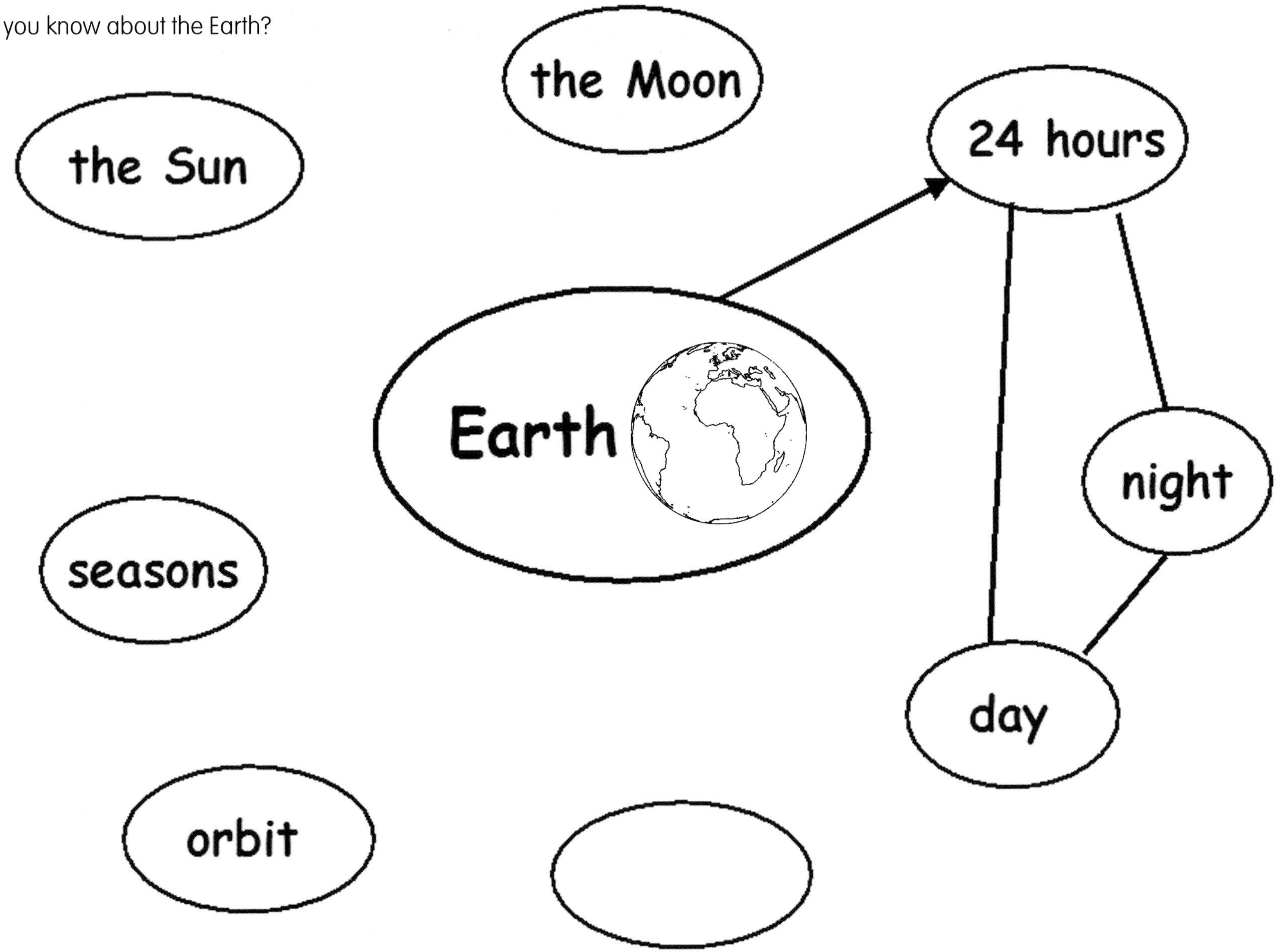

- Add other words and lines to join them.

Time zones

A group of scientists met in Washington, USA, in 1884 to work out the Earth's time zones.

They marked a globe in circles known as **lines of longitude**:

Side view

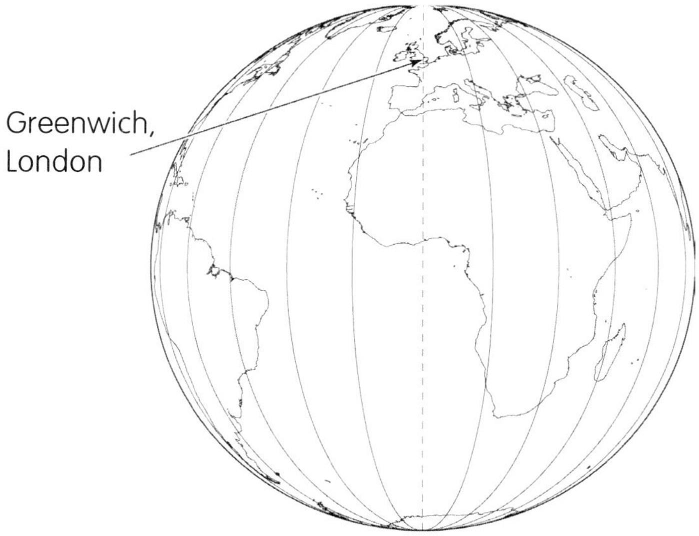

View from North

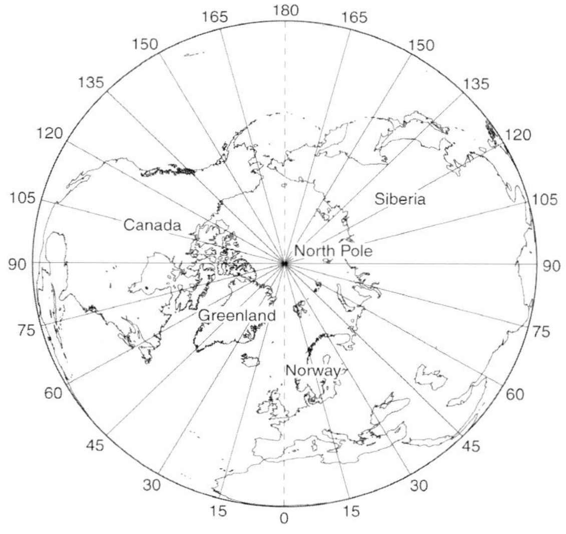

They marked the lines in degrees. 360° = a complete circle.

The Earth spins through 360° in a complete rotation.

How long does this take? ____________________

How long does it take to turn through 15°? ____________________

- Find the longitude of these cities.
  What time will it be in each city?

City	Longitude	Time
Accra, Ghana		
Canberra, Australia		
Durban, South Africa		
Rio de Janeiro, Brazil		
Rome, Italy		

Photocopiable sheet 13

Sound vibrations

Try these activities.

- Record what you see and hear.
- Explain your observations.

Strike a tuning fork on a thick book. Move it along a piece of paper. 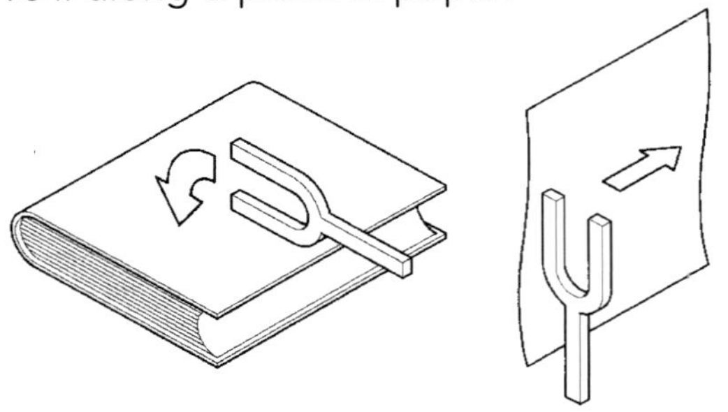 I saw ____ I heard ____ Explanation ____ ____	Strike a tuning fork on a thick book. Dip it into a saucer of water. 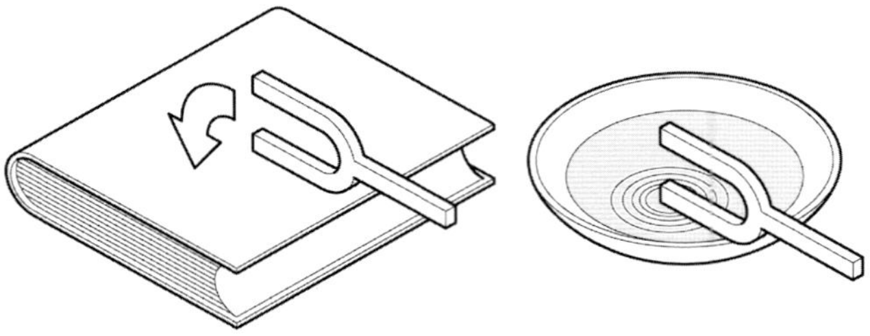 I saw ____ I heard ____ Explanation ____ ____
Strike a tuning fork on a thick book. Hold it close to a table tennis ball. 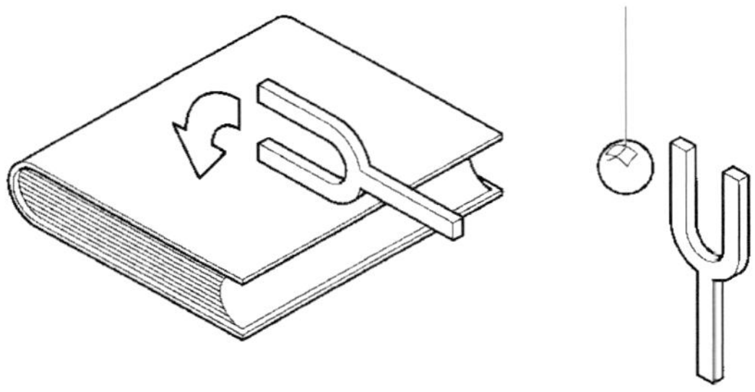 I saw ____ I heard ____ Explanation ____ ____	Hold a ruler over the edge of a table using a thick book. Twang the ruler. 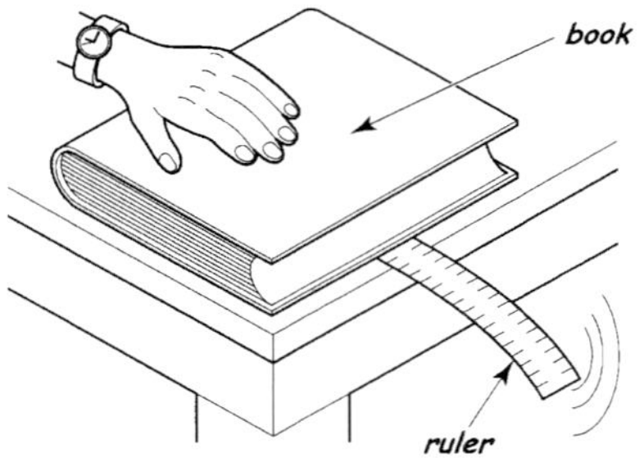 I saw ____ I heard ____ Explanation ____ ____

What have you learned about sounds? ____

____

____

Resources

Keeping healthy
British Potato Council http://www.potatoesforschools.org.uk/
British Nutrition Foundation http://www.nutrition.org.uk/information.html
Healthy Eating Programme, Bangor Food Research Unit, University of Wales Bangor http://www.fooddudes.co.uk/ (Includes activities for children)
Alliance for Better Food and Farming http://www.sustainweb.org/grab5_index.asp
Cool Meals http://www.soolmeals.co.uk/ (Includes activities for children)
Kids' Health http://www.kidshealth.org/kid/stay_healthy/ (Includes activities for children)
Scurvy and vitamin C http://www.bbc.co.uk/history/discovery/exploration/captaincook_scurvy_01.shtml,
http://www.blupete.con.Hist/Gloss/Scurvy/htm,
http://pc-78-120,udac,se:8001/WWW/Nautica/Medicine/Lind(1753).html,
http://www.mc.vanderbilt.edu/biolib/he/journeys/book6.html.

Life cycles
The Usborne Nature Trail Omnibus, Malcolm Hart, Sue Tarsky, Ingrid Selberg, Su Swallow & Ruth Thomson, 1990, Usborne
Poisonous plants: http://www.rbgkew.org.uk/data/poisplts.html

Gases around us

Solid
It can be piled up.
It keeps its shape when taken out of a container.
It cannot be poured unless it is in many pieces.
It cannot flow through tubes or pipes.
If you put it on a level surface such as a tabletop, it should not run off.
If you make a hole in the bottom of a container of solid, the solid stays inside unless it is in small pieces which can fit through the hole.
If you open the top of a container of solid, the solid stays inside it.
If you tip it out of a container, it falls downwards.
Liquid
It cannot be piled up. It does not keep its shape unless it is in a container.
It can be poured.
It can flow through tubes or pipes.
If you put it on a level surface such as a tabletop, it runs off.
If you make a hole in the bottom of a container of liquid, the liquid runs downwards.
If you open the top of a container of liquid, the liquid stays inside it.
If you tip it out of a container, it falls downwards.
Gas
It cannot be piled up.
It does not keep its shape when taken out of a container.
It cannot be poured.
It can flow through tubes or pipes.
You cannot put it on a level surface such as a tabletop because it escapes in all directions.
If you make a hole in the bottom of a container of gas, the gas escapes in all directions.
If you open the top of a container of gas, the gas escapes in all directions.
If you let it out of a container it goes in all directions.

How a spray bottle works http://www.sience.howstuffworks.com/question673.htm

Changing state
Weather *How Weather Works: Understanding the Elements* (Rene Chaboud, Thames & Hudson)
http://www.kean/edu/~gkolodiy/physics/newton/#Friction%20Sim%201Info (a simulation of how friction affects movement)
http://www-estp.gsfc.nasa.gov/stargaze/Snewton.htm (Isaac Newton)

Resources

Earth, Sun and Moon
Finding Out About Sun, Moon and Planets (Lynn Myring & Sheila Snowden, Usborne)
NASA (activities and research for children) http://www.kids.msfc.nasa.gov/
BBC science (activities and information for children, including an interactive simulation of the rotation of the Earth and the Moon, and the Earth's orbit around the Sun) http://www.bbc.co.uk/schools/scienceclips/ages/9_10/earth_sun_moon.shtml
Moon phases http://www.home.hiwaay.net/~krcool/Astro/moon/moonphse/
Information about the Moon (for teachers) http://www.inkido.indiana.edu/a100/earthmoon2.html

Changing sounds
BBC Schools science clips (interactive activities for children)
http://www.bbc.co.uk/schools/scienceclips/ages9_10/changing_sounds.shtml
Kids' Health (the ear and hearing) http://www.kidshealth.org/kid/body/ear_noSW.html7
Learn.co.uk (includes the poem The Sound Collector by Roger McGough and some very simple online activities – suitable for lower-attaining children) http://www.learn.co.uk/glearning/primary/lessons/ks1/sounds3/default.asp
National Institute on Deafness and other Communication Disorders (an interactive 'sound ruler' showing decibel levels of sounds from 0 to 140 decibels) http://www.nidcd.nih.gov/health/education/decibel.asp
Royal National Institute for the Blind (the life of Helen Keller)
http://www.rnib.org.uk/xpedio/groups/public/dicuments/PublicWebsite/public_keller.hcsp
REPO (How Alexander Graham Bell developed the telephone) http://www.iath.virginia.edu/albell/homepage.html

Morse code

A .-	**G** --.	**S** ...	**0** -----	**5**	**Full stop** .-.-.-
B -...	**H**	**T** -	**1** .----	**6** -....	**Comma** --..--
C -.-.	**I** ..	**U** ..-	**2** ..---	**7** --...	**Question mark** ..--..
D -..	**J** .---	**V** ...-	**3** ...--	**8** ---..	
E .	**N** -.	**W** .--	**4** ....-	**9** ----.	
F ..-.	**O** ---	**X** -..-			
K -.-	**P** .--.	**Y** -.--			
L .-..	**Q** --.-	**Z** --..			
M --	**R** .-.				

For translating text into Morse code
http://www.scphillips.com/morse/index.html?
http://www.scphillips.com/morse/trans.html